Mohamed Abdel-Raheem
Mariam Hussain Chaffat

Entomopathogenic fungi as the most promising biological control agents

Mohamed Abdel-Raheem
Mariam Hussain Chaffat

Entomopathogenic fungi as the most promising biological control agents

Entomopathogenic fungi as biological control agents

Noor Publishing

Imprint

Cover image: www.ingimage.com

Publisher:
Noor Publishing
is a trademark of
Dodo Books Indian Ocean Ltd. and OmniScriptum S.R.L publishing group

120 High Road, East Finchley, London, N2 9ED, United Kingdom
Str. Armeneasca 28/1, office 1, Chisinau MD-2012, Republic of Moldova, Europe
Managing Directors: Ieva Konstantinova, Victoria Ursu
info@omniscriptum.com

Printed at: see last page
ISBN: 978-620-0-04543-0

Entomopathogenic fungi as the most promising biological control agents

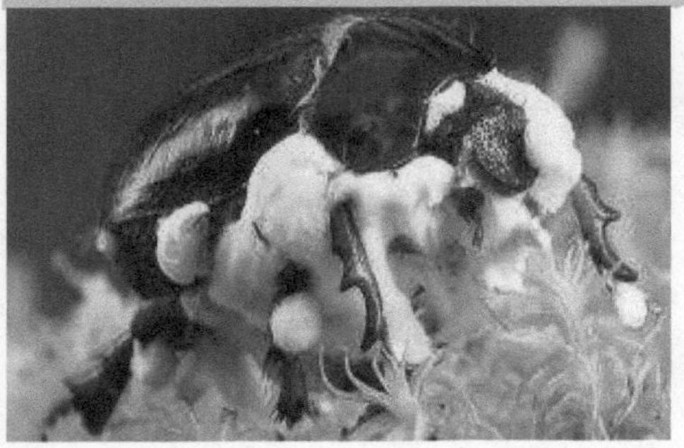

By

Prof. Dr. Mohamed Abdel-Raheem

Assistant teacher. Mariam Hussain Chaffat

(2025)

Entomopathogenic fungi as the most promising biological control agents

Prof. Dr. Mohamed Abdel-Raheem Ali Abdel-Raheem

Professor of Entomology (Biological Control)

Pests & Plant Protection Department

Agricultural and Biological Research Institute

National Research Centre.

Assistant teacher.Mariam Hussain Chaffat

Department of plant protection, College of Agriculture , University of Kerbala, Iraq

Content list

Introduction

Agricultural pests include plant pathogens (Entomopathogenic fungi). There are estimated to be about 67,000 different pest species.

They are constraint on agricultural production, responsible for around 40% loss of potential global crop yields. These losses occur despite the considerable efforts made at pest control, and they suggest that control on pest management are the best way forward for improving yields and access to food.

Pest problems are inevitable part of agriculture. They occur seriously because systems of agriculture are very simplified, not stable modifications of natural ecosystems. The agricultural land disrupts the ecological forces that regulate potential insect pests species in natural ecosystems.

Farming can do the physico-chemical environment more favourable for pest activity, for example irrigation or the warm conditions that found in glasshouses. Finally, using of pesticides will destroy natural predators and parasites that help keep pests always under control.

The most important problems happened when the pests develop resistance to insect pesticides. Now, over 500 arthropod species now show resistance to one or more types insect pesticides. Other dangerous problems are occurred by alien species that are accidentally introduced to a new country or continent and which escape their evolved natural predators More than 11,000 alien species.

These cause the greatest damage to crops. There are also threats from emerging pests. Farmers are under pressure to reduce the using of insect pesticides without loss of yields or crop quality, but the controlling of pests is becoming increasingly very difficult due to insect pesticide resistance and the decreasing availability of products.

Biological control is using by man of a living organisms to help manage the population density of a pest organism. Natural enemies are organisms that kill or debilitate another organism. Biological control programmes show throughout the world in agriculture. Biological control agents include the following:

(1) Predatory insects and mites, which eat their prey

(2) parasitoids, which are insects with a free living adult stage and a larval stage that is parasitic on another insect

(3) Parasites and microbial pathogens, such as nematodes, fungi, bacteria, viruses and protozoa, which cause lethal infections.

In Europe, typical examples of where these agents are being used include:

(1) The application of parasitoids to control whiteflies in glasshouses

(2) parasitic nematodes against slugs

(3) mycoparasitic fungi to control plant diseases of horticultural crops

(4) Use of viruses to control codling moth in apple orchards; and building habitats on farms to increase natural populations of predators and other beneficial organisms.

Natural enemies represent a large component of the world's biodiversity, for example parasitoids account for about 10% of the world's species, and each agricultural pest species is fed upon by up to hundreds of species of natural enemies. However, only a small proportion of the available species have been investigated for crop protection.

Thus, while 750 species of insect pathogenic fungi are known, less than 20 have received serious attention as control agents of insect pests.

The natural enemies inhabiting an agro ecosystem play a key role in preventing pests reaching damaging levels. The ways in which biological control agents are used vary according to the type of pest, the biological characteristics of the control agent, as well as the agricultural setting.

People who use biocontrol are bound by the FAO code of conduct on the import and release of biological control agents as well as national legislation on releases into the environment.

The use of microbial agents is governed by Plant Protection Product regulations, in which authorisations proceed along similar lines to those for chemical pesticides although there are also some specific schemes. BCAs have a range of attractive properties that include host specificity, lack of toxic residue, no phytotoxic effects, human safety, and the potential for pest management to be self sustaining. Many are able to actively locate their prey.

BCAs can also be produced locally which can be important in terms of choosing and matching natural enemies to small scale needs. Successful use requires fundamental knowledge of the ecology of both the natural enemy and the pest. When this condition is satisfied, and the agent is used firmly within IPM, then biological control can sometimes be more cost effective than purely chemical control.

However, under the present market system, many biological control products have not competed well with less expensive and more effective synthetic pesticides. The down sides of BCAs are that most are niche products, pest control is not immediate, there can be lack of environmental persistence, and efficacy can be low and unpredictable particularly in outdoor environments. The approvals process used for microbial agents results in significantly greater authorisation costs than for macro- agents.

Entomopathogens Many farmers and growers are now familiar with the use of predators and parasitoids for biological control of arthropod (insect and mite) pests, but it is also possible to use specific micro-organisms that kill arthropods. These include entomopathogenic fungi, nematodes, bacteria and viruses. These are all widespread in the natural environment and cause infections in many pest species.

Entomopathogens contribute to the natural regulation of many populations of arthropods. Much of the research in this area concerns the causal agents of insect diseases and their exploitation for biological pest control. Many entomopathogens can be mass produced, formulated, and applied to pest populations in a manner analogous to chemical pesticides, i.e. as nonpersistent remedial treatments that are released inundatively.

Entomopathogens have also been used as classical biological control agents of alien insect pests, and natural pest control by entomopathogens has been enhanced by habitat manipulation. Entomopathogenic fungi. There are thought to be about 750 species of fungi that cause infections in insects or mites. As a group, they attack a wide range of insect and mite species, but individual species and strains of fungus are very specific.

The fungi produce spores which infect their host by germinating on its surface and then growing into its body. Death takes between 4 and 10 days, depending on the type of fungus and the number of infecting spores. After death, the fungus produces thousands of new spores on the dead body, which disperse and continue their life cycle on new hosts. There are two main taxonomic orders of entomopathogenic fungi.

The Entomophthorales occur in the phylum Zygomycota and include genera such as Pandora, Entomophthora and Conidiobolus. Many of the species are co-evolved, obligate pathogens that show specific ecomorphological adaptations to the life cycles of their hosts, such as the production of forcibly-ejected infective spores that are produced on insect cadavers during the night, when environmental conditions are most conducive to infection.

These fungi often cause natural epizootics in insect and mite populations. However, some of them are very difficult to mass produce in culture, which is a challenge for people wanting to develop them as biopesticides. Despite this, they can be used for pest control by being applied to the field (as mycelium or as infected insect cadavers). For example, isolates of Neozygites floridana from S. America are being investigated for classical biological control of the cassava green mite Mononychellus tanajoa, a major alien pest of cassava in Africa.

Elsewhere, natural epizootics of entomophthoralean fungi are being used as a natural form of pest control. For example, a prediction service developed by Don Steinkraus at the University of Arkansas is used to inform cotton growers in the southeast USA about outbreaks of Neozygites fresenii in cotton aphids, Aphis gossypii. These outbreaks can reduce aphid populations by about 80% in 4 days and can be predicted up to ten days in advance. Cotton farmers then have the choice to withhold expensive insecticide sprays.

The second major order of entomopathogenic fungi – the Hypocreales – occurs in the phylum Ascomycotina. There are many species in the Ascomycotina in which the sexual phase (teleomorph) is not known and which reproduce entirely asexually, and which have only be confirmed to be members of the Ascomycota by using molecular (DNA) data. These fungi are commonly known as anamorphic fungi. Sexually reproducing hypocrealean fungi occur in the genera Cordyceps and Torrubiella. Cordyceps is thought to originate in Asia and Cordyceps species are important natural parasites of many insect and mite species in tropical forests. Cordyceps sinensis is a pathogen of ghost moth (Thitarodes) caterpillars that occur at high altitude in the Himalayas, Tibet and China. In China it is referred to as dong chong xia cao (winter animal summer grass).

It is a highly valued herbal medicine and contains a range of pharmalogically active compounds. Important anamorphic genera of entomopathogenic fungi include Beauveria, Isaria, Metarhizium and Lecanicillium. Species from all these genera are used as biopesticides of insect pests. Recently, molecular methods have revealed the telemorph connections of these fungi. Molecular studies are also shedding light on the phylogenetic relationships with other fungi. Much of this work is being pioneered in the US by mycologists such as Meredith Blackwell, Richard Humber, Steve Rehner and Joey Spatafora.

They have shown that entomopathogenicity has evolved multiple times in the ascomycete fungi, characterised by host shifting from an intimate association with plants to insects and vice versa. The spores of many species of the anamorphic entomopathogenic fungi can be mass produced on a variety of culture media, and so are suitable for development as biopesticides which are applied inundatively to pest populations.

They have a range of desirable characteristics including safety to people, compatibility with other natural enemies, and a lack of toxic residues. They also offer the possibility of providing persistent control by multiplying in the pest population. Because they have contact action, they are good for the control of sap feeding pests, like aphids and whiteflies, which cannot be infected by other types of biopesticide (such as bacteria and viruses) which are active only when ingested. The downside is that they are more expensive than chemicals and can be affected adversely by certain environmental conditions.

There can also be a false expectation that biopesticides will have the same efficacy as chemical insecticides, whilst their desirable biological qualities are overlooked. In addition, the costs of registration are high in relation to the current market size, deterring companies from pursuing registration. At present there are two commercial products are available to UK growers, namely Vertalec (for control of glasshouse aphids) and Mycotal (for glasshouse whitefly).

Both are based on species of the fungus Lecanicillium: Vertalec is based on Lecanicillium longisporum, and Mycotal is based on Lecanicillium muscarium (until recently, these were considered to be one species, Verticillium lecanii, but the genus Verticillium has now been revised). They are produced by Koppert BV, and the products were originally developed at the Glasshouse Crops Research Institute, a forerunner of Warwick HRI. Products based on Beauveria bassiana and Metarhizium anisopliae are expected to be available shortly in the UK for the control of glasshouse pests and vine weevils

Biological control is a component of an integrated pest management strategy. It is defined as the reduction of pest populations by **Pathogens**.

Pathogens are disease-causing organisms including **bacteria, fungi, and viruses**.

They kill or debilitate their host and are relatively specific to certain insect groups.

Fungus-infected adult flies are often common following periods of high humidity.

Microbial pesticides

The microorganisms like **virus, fungus, protozoan**, or **bacterium** are the active ingredient in this type of pesticides. Each microorganism have specific active ingredient to control the specific pests, but microbial pesticides can control or kill many kinds of pests which damage the crop production. One fungus can control the weeds and other control or kill the insects-pests.

Biopesticide characteristics

Biopesticides have some characteristics:

1. Target range of biopesticides is narrow and used for specific problem.
2. Biopesticides work slowly.
3. Biopesticides suppress the pest population rather than eliminate.
4. Application time of biopesticides is somewhat critical.
5. There is limited field persistence and shelf life.
6. Biopesticides often used as amount of IPM programs.
7. They are much more suitable for human as well as for environment than conventional pesticides.
8. No residue problems commonly present by biopesticide.

Biopesticide application on food safety

Food safety requires wide application range of biopesticides which includes:

If one organism is feed on other organism, this is called its natural enemy. The beneficial insects are those insects that feed on pests. In order arthropods, for example, mites and spiders are beneficial.

Microbial pesticide exists naturally or by genetically changed fungi, protozoans, bacteria, algae, or fungi. This can be used as an alternate method to chemical insecticide, this is very effective.

Biological toxin material is derived from microorganism, for example, fungus or bacterium, this is called microbial toxin. These types of microorganism may cause death or rupture the gut of the pest because these entomopathogens are highly toxic. Studies proved that pathogen develops insecticidal toxin that are very much important in pathogenesis.

Antibiotics are the substance that stops the growth or kills microorganism, including both fungi and bacteria.

In "bactericidal", the antibiotics kill the bacteria, whereas in "bacteriostatic", the antibiotics stop the growth bacteria. It is the success of biotechnology that develops the transgenic crops which are resistance against the major pests and also commercialize the transgenic crops. In first generation, the products include plants with just single insecticidal *Bt* genes, which shows resistance against the major pests of cotton and corn.

In the search for alternative solutions to crop protection problems, the interest in plants and their chemo-biodiversity as a source of bioactive substances has increased. Plants are capable of synthesizing an overwhelming variety of small organic molecules called secondary metabolites, usually with very complex and unique carbon skeleton structures.

For many years for the protection of crops, these substances are used for the benefit of mankind.

Plants release some chemicals in environment and when they are used as smother crops, green manures, intercrop, cover crop or mulch, or in rotational sequences grown than it can invade insect-pests and pathogenic diseases and shows the progress in the yield of crops.

Botanicals include crude or semi refined extracts and isolated or purified compounds from various plant species and commercial products.

Microbial pesticides

Entomopathogenic fungi are considered as essential natural regulators in population of different insects. These fungi have potential as mycoinsecticide agents alongside different agricultural insect pests.

The mode of actions of these fungi mainly cause infection in body of their host through cuticle penetration, acquisition entrance to the host hemolymph, production of toxins, and grow by consuming nutrients which are existing in the hemocoel to avoid immune responses in insects.

A total of 1600 different types of viruses are involved in the infection of 1100 different species of insects and mites.

Baculovirus is considered as a distinct group of viruses. About 100 species of insects are susceptible to this group of viruses. The main feature of this virus group is host specification.

Bacterial biopesticides are considered as the most common and inexpensive method of microbial pesticides. Different species of butterflies and moths as well as species of flies, beetles, and mosquitoes are generally controlled by these biopesticides. Bacteria which are used as bacterial pathogens to manage insect pest populations are mainly spore-forming and rod-shaped bacteria in the genus *Bacillus*.

The EPNs in introduces symbiotic, pathogenic bacteria of the genera *Xenorhabdus* (in the Steinernematidae) or *Photorhabdus* (Heterorhabditidae) into the hemocoel of their hosts following penetration. Subsequent multiplication of the bacteria leads to host death, which can occur within as little as 48 hours of infection.

Nematodes infected insects usually have abnormal behavior than uninfected individuals. The nematode can kill its host without its associated bacterium but is unable to reproduce without it.

They have also been used with commercial success against citrus root weevil, against turf pests and on mushroom crops (Figure 1).

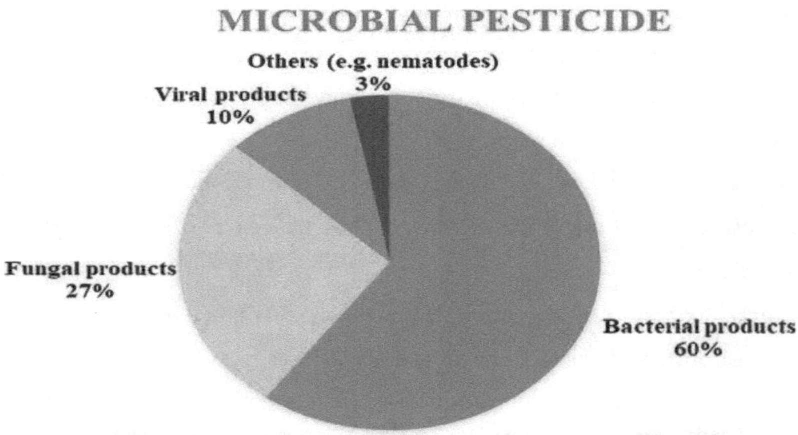

Microbial biopesticide products worldwide

Microbial biopesticide product worldwide active ingredient: microorganism (bacteria, fungi, virus, nematodes, etc.) (Division of Agricultural Chemicals, 2011).

Entomopathogen groups

Important entomopathogen groups and the modes of their infection process are described below.

Entomopathogenic fungi have played a uniquely important role in the history of microbial control of insects. Historical evidence indicated that entomopathogenic fungi were the first to be recognized as microbial disease in insects [1].

Beauveria bassiana, commonly known as white muscardine fungus attacks a wide range of immature and adult insects. *Metarhizium anisopliae* a green muscardine fungus is reported to infect 200 species of insects and arthropods. Both of these entomopathogenic fungi are soil borne and widely distributed. These fungi have been documented to occur naturally in over 750 species of host insects [2-4].

Soil is considered to be the natural environment of these fungi because they deposit their infectious spores there and remain in the soil for some duration of their life cycle. Therefore, it was determined that soil is the most appropriate place to determine their occurrence [5]. The occurrence and distribution of insect pathogenic fungi in agricultural field soils have been extensively investigated in previous studies [6-11].

Nevertheless there are studies on isolation of these fungi from insect cadavers as well [12-17].

Wild isolates are still of great importance due to their potential unique characteristics in biological control of insect pests. The presence of certain entomopathogenic fungal species can be considered as an indicator of their ability to survive in that environment. This information is useful for the selection of biocontrol agents because the indigenous dominant species are generally the most suitable candidates [18].

Factors such as the geographical location, climate, habitat, altitude, and pH of the soil or organic matter impact the presence of fungal species, and the response of each species to these conditions varies [18]. Use of *Galleria*-bait technique is a common method for isolation of entomopathogenic fungi from soil [19-22].

Although identification of the common entomopathogenic fungi like (*Beauveria bassiana, Metarhizium anisopliae,* and *Verticillium lecanii*) looks easy even to non-mycologists because they have basic diagnostic characters making them easily identifiable.

It must be remembered that these species have other complex molecular, morphological and pathobiological traits [23, 24].

Entomopathogenic fungi typically cause infection when spores come in contact with the arthropod host. Under ideal conditions of moderate temperatures and high relative humidity, fungal spores germinate and breach the insect cuticle through enzymatic degradation and mechanical pressure to gain entry into the insect body.

Once inside the body, the fungi multiply, invade the insect tissues, emerge from the dead insect, and produce more spores. Natural epizootics of entomophthoralean fungi such as *Entomophaga maimaiga* (in gypsy moth), *Entomophthora muscae* (in flies), *Neozygites fresenii* (in aphids), *N. floridana* (in mites), and *Pandora neoaphidis* (in aphids) are known to cause significant reductions in host populations.

Although these fastidious fungi are difficult to culture in artificial media and do not have the potential to be sold as biopesticides they are still important in natural control of some pest species.

Hypoclealean fungi such as *Beauveria bassiana*, *Isaria fumosorosea*, *Hirsutella thompsonii*, *Lecanicillium lecanii, Metarhizium acridum, M. anisopliae*, and *M. brunneum*, on the other hand, are commercially sold as biopesticides in multiple formulations around the world. Fungal pathogens have a broad host range and are especially suitable for controlling pests that have piercing and sucking mouthparts because spores do not have to be ingested.

However, entomopathogenic fungi are also effective against a variety of pests such as wireworms and borers that have chewing mouthparts.

Related to fungi, the spore-forming microsporidium, *Paranosema* (*Nosema*) *locustae* is a pathogen that has been used for controlling locusts, grasshoppers, and some crickets.

When *P. locustae* is ingested, the midgut tissues become infected, followed by infection in the fat body tissues. The disease weakens and eventually kills the orthopteran host within a few weeks.

There are fungus species that can infect and kill insects. These fungi are called insect pathogenic fungi or entomopathogenic fungi.

These fungi are very specific to insects, often to particular insect species, and do not infect animals or plants.

Most insect- pathogenic fungi need humid conditions for infection and development but some fungus species can also infect insects when it is dry.

Free help from Mother Nature!

When an insect has been killed by a fungus, the fungus emerges from the insect's body to produce spores that can sometimes be seen as a "dusty" or "cottony" appearance.

When you find many insects dead and covered with fungus, you can consider making your own bio-insecticide from these dead insects.

Collect as many as you can find in the field, put them in a jar with water, crush them a little and stir well.

This will release fungus spores into the water. Filter the water slightly to remove large insect parts. The remaining solution can be used to test

Its Spray the solution over insects that are placed in a jar. Check if these insects become infected over the next days. Spray another jar of insects with pure water as a control.

Beauveria bassiana *Metarhizium anisopliae*

Fig. (1): Insects infected with Entomopathogenic Fungi

Mode of action of entomopathogenic fungi:

With most fungi disease development can be divided into nine steps as given by **Roberts,** [25]:

1. Attachment of the infective unit (e.g. conidium or zoospore) to the insect epicuticle.
2. Germination of the infective unit on the cuticle.
3. Penetration of the cuticle, either directly by germ tubes or by infection pegs from appressoria.
4. Multiplication in the yeast phase (hyphal bodies) in the haemocoel.
5. Production of toxic metabolites.
6. Death of the host.
7. Growth in the mycelial phase with invasion of virtually all host organs.
8. Penetration of hyphae from the interior through the cuticle to the exterior of the insect.
9. Production of new infective units (conidia).

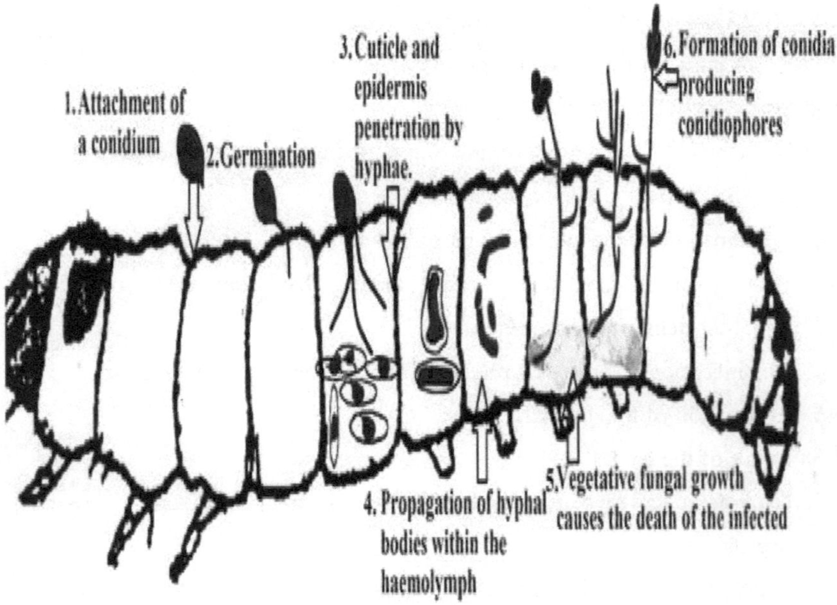

Fig. (2): Mode of action of Entomopathogenic Fungi

Isolation of Entomopathogenic fungi

Isolation from soil

Biweekly soil samples were collected from three fruit orchards from area aroud Alexandria desert road starting from January to June 2015. A total of 100 soil samples were collected throughout the season.

These orchards contained trees of date palms, olives, mangoes, annonas (keshta fruit), pomegranates, pears, grapes and other fruit species. The trees were mixed in orchards and they were 7 years or older. The soil was sandy with organic material transferred from cattle barns.

The entomopathogenic fungi were isolated from soil samples using the *Galleria*-bait technique based on [25]. In this technique, soil samples were collected from the soil at 0-20 cm-depth in a circle of 1m-diameter around tree trunks. A weight of approximately 1.5kg of soil was collected in plastic bags and transferred in a cooled box to the laboratory. In the laboratory the samples were divided into 3 parts in plastic containers. Larvae of the greater wax moth *Galleria mellonella* were distributed in the cups at a rate of 3larvae/ cup and wetted to approximately 15% water content.

After 7days of incubation at 25°C, infected larvae were transferred to 5cm-diameter Petri dishes furnished with wetted filter paper and left for development and sporulation. Numbers of positive samples (samples containing entomopathogenic fungi) and numbers of infected *G. mellonella* larvae in each sample were recorded, [26].

Identification and maintenance of obtained fungi

Extracted entomopathogenic fungi were identified morphologically based on the morphological characteristics of reproductive structures according to [26, 27].

The obtained isolates could be identified as *Beauveria bassiana, Metarhizium anisopliae* and *Verticillium lecanii*. Natural abundance for populations of the three entomopathogenic fungi in studied fruit orchards is summarized in Table 1 as percentages of positive samples and percentages of infected *G. mellonella* larvae on every sampling date.

The white muscardine fungus *B. bassiana* was the highest abundant forming 25.03% of total collected samples followed by the green muscardine fungus, *M. anisopliae* (17.76%) then the metallic pink fungus *V. lecanii* (14.49). *B. bassiana* occurred from the first of January until the middle of April.

Percentages of positive samples containing this fungus in each sampling date ranged from 10 to 70% with a peak of occurrence in early January and a minimum in mid April.

Percentages of infected *G. mellonella* larvae in collected samples ranged from 1.11 to 17.78%. Our results showed high occurrence of entomopathogenic fungi in soils sampled from fruit orchards.

In a survey study conducted by [26], positive soil samples containing *B. bassiana* or *M. anisopliae* comprised less than 2% of total samples collected from El Behaira, Kafr Elshaikh (north of Egypt) or Aswan (south of Egypt). Like what we found in this study, *B. bassiana* was the most frequent entomopathogenic fungus in the soils followed by *M. anisopliae* in Spain [25].

In China, *B. bassiana* was more abundant than *M. anisooliae* in soils [26].

Table 1: Percentages of positive soil samples collected from fruit orchards around Alexandria desert road and containing the entomopathogenic fungi *Beauveria bassiana, Metarhizium anisopliae* and *Verticillium lecanii* in relation to the sampling date, [26].

Sampling dates	*Beauveria bassiana*		*Metarhizium anisopliae*		*Verticillium lecanii*	
	Positive samples	Infected larvae	Positive samples	Infected larvae	Positive samples	Infected larvae
1.1.15	70	14.44	10	4.44	0.0	0.0
15.1.15	40	17.78	50	6.67	30	12.22
1.2.15	60	16.67	20	2.22	10	3.33
15.2.15	30	3.33	10	1.11	10	1.11
1.3.15	20	2.22	0.0	0.0	40	5.56
15.3.15	10	1.11	0.0	0.0	0.0	0.0
1.4.15	0.0	0.00	0.0	0.0	0.0	0.0
15.4.15	30	3.33	40	4.44	0.0	0.0
1.5.15	0.0	0.0	10	3.33	0.0	0.0
1.6.15	0.0	0.0	0.0	0.0	0.0	0.0
Mean	25.03	7.33	17.76	2.40	14.49	3.99

M. anisopliae had two periods of occurrence; the first was from early January to early March while the second was from mid April to early May with a peak of 50% positive samples and a minimum of 10%. Percentages of *G. mellonella* larvae that infected with *M. anisopliae* in collected soil samples were between 1.1 and 6.67%. *V. lecanii* occurred from mid January to early March recording 10-40% positive samples and 1-12.2% infected *G. mellonella* larvae, [43].

Relationship between fungus occurrence and plant cover

Percentages of soil samples containing different entomopathogenic fungi are given in Fig 1. *B. bassiana* was found in soil under all kinds of studied fruit trees recording 10-40 % of collected samples. This fungus was found mostly under Mango trees (50% of samples) then under Pomegranates (40%). Its occurrence was between 10-30% for the rest of fruit kinds. *M. anisopliae* was found under seven fruit kinds. Its distribution looked almost even among the fruit kinds. Its range of occurrence was between 10 and 30% of collected samples. *V. lecanii* was found under 6 kinds of fruits with no clear relationship between the fungus type and the fruit kind, [26].

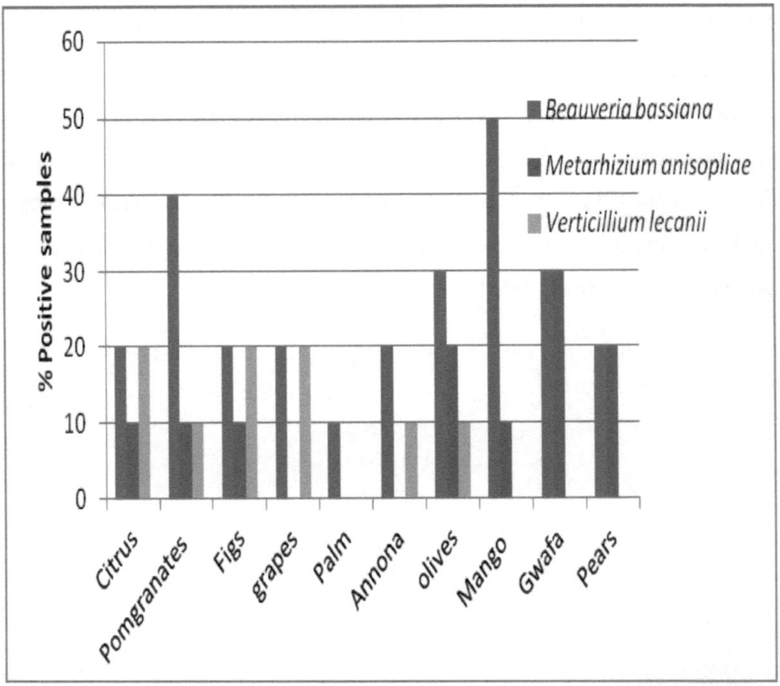

Fig 3: Abundance of *Beauveria bassiana, Metarhizium anisopliae* and *Verticillium lecanii* in orchards around Alexandria desert road in relation to the kind of fruit, [26].

No clear relationship between the distribution of entomopathogenic fungi and the kind of fruit tree could be determined. Olives, mangos, pomegranates and guava supported relatively higher rates of fungus occurrence (60%) compared to the other kinds of fruits. Palms however had only 10% of fungal occurrence, [31].

We suggest that fungal distribution among fruit trees depended on the fungus species and the rate of organic fertilization under the trees rather than the kind of the tree. These kinds of trees usually receive more manure than the others. Farmers in Egypt do not usually fertilize palm trees, [25]. Charnley, [26] mentioned the organic content and temperature among the factors affecting fungal abundance and activity in the soil.

Isolation from Insects cadavers:

Collected cadavers from the field in Kafr El-Sheikh Governorate and kept in Petri- dishes (9 cm diameter x 1.5 cm) at 24 °c. Incubated dishes were inspected daily to observe the fungal growth that was purified and used to confirm the disease cycle (Koh's postulates), then stored on slant of PDA artificial media at 4 °c until used in subsequent experiments, [12].

Isolation of new fungal generation from the infected insects:

The infected insects which covered with fungal mycelium were collected from the field and carried to the laboratory.

Different stages from *Cassida vittata* and *Scrobipalpa ocellatella* were found died and totally covered with fungal mycelium. The insect cadavers were put on a wetted filter paper in a Petri – dish and incubated at 24 ± 1 ° c for seven days. The new fungal generation was isolated from the surface of insect cadaver and cultured on PDA medium in Petri-ditches. Fungal cultured were purified weekly until pure cultures were obtained, [12].

Two isolates of *Beauveria bassiana* and *Metarhizium anisopliae* were isolated from infected cadavers of *Cassida vittata* and *Scrobipalpa ocellatella* from two regions (Beila and Abu Kalab) in Kafr El-Sheikh Governorate as given in table (2) [12].

Table (2): The entomopathogenic fungi isolated from Insect cadavers from different Regions in Kafr El- Sheikh during February 2000 to May 2003[12]:

Fungi	Original host	Host plant	Location	Date
B. bassiana	*S. ocellatella* (4 th larvae)	Sugar -beet	Abu kalab	May 2001
B. Bassiana	*C. vittata* (Larvae-adults)	Sugar- beet	Abu Kalab	April 2001
M. anisopliae	*C. vittata* (Larvae-adults)	Sugar- beet	Beila	April 2003

Table (3): Numbers of positive entomopathogenic fungi recovery by trapping with larvae of *G. mellonella* from soil samples collected from three governorates and their rates of incidence (%), [26].

Locality	Number of samples	Fungi positive samples and incidence (%)	
		B. bassiana	*M. anisopliae*
Elbehaira	119	3 (2.52%)	1 (0.84%)
Kafr El-Shaeikh	103	2 (1.94%)	2 (1.94%)
Aswan	97	1 (1.03%)	1 (1.03%)
Total	319	6 (1.88%)	4 (1.25%)

Mass Production of Entomopathogenic fungi

Fungi were grown and maintained on:

1- peptone medium (10g Peptone, 40g Dextrose, 2g yeast extract 15g Agar and 500 ml chloramphenicol and completed to one liter by distilled water). The medium was autoclaved at 120°c for 20 minutes, and poured in Petri-dishes (9 cm diameter x 1.5 cm) then inoculated with the entomopathogenic fungi and kept at 25 ±2°C and 85 ±5 RH. The fungal isolates were re-cultured every 15-30 days and kept at 4°C.

2- Potato dextrose agar (PDA) (250 g potatoes, 20 g agar and 1000 ml distilled water. The media was autoclaved at 120 °c for 20 minutes, and poured in Petri- dishes (9 cm diameter x 1.5 cm). The fungal isolates were re-cultured every 14-30 days and isolates were kept at 4 °c. To restore the virulence of the isolates they were passed through their natural host or through the wax moth larvae *Galleria mellonella*.

3- *B. bassiana* and *M. anisopliae* isolates were propagated on wetted rice. Two Kilos gram wetted rice was washed in boiled water for 10 min. and put in thermal bags. These bags were autoclaved at 120°C for 20 min., then infected by isolates and incubated at 25 ±2°C for 15 days. The Conidia were harvested by distilled water and filtered through cheese cloth to reduce mycelium clumps and Tween 80% was added, **Lacey,** [27].

These are some of the most common insect pathogenic fungi:

Beauveria bassiana

***Beauveria bassiana* is** a fungus that grows naturally in soils throughout the world and acts as a parasite on various arthropod species, causing white muscardine disease; it thus belongs to the entomopathogenic fungi .

It is being used as a biological being used as a biological insecticide to control a number of pests such as termites ،thrips ،whiteflies ،aphids and different beetles.

Preferred Scientific Name

- *Beauveria bassiana* (Bals.-Criv.) Vuill. (1912)

Preferred Common Name

- white muscardine fungus

Other Scientific Names

- *Beauveria brongniartii* subsp. delacroixii Sacc. (1892)
- *Beauveria densa* (Link) F. Picard (1914)
- *Beauveria doryphorae* R. Poiss. & Patay (1935)
- *Beauveria effusa* (Beauverie) Vuill. (1912)
- *Beauveria laxa* Petch (1931)
- *Beauveria paradoxa* (Bals.-Criv.) Ramsb. (1941)
- *Beauveria shiotae* (Kuru) Langeron (1936)
- *Beauveria stephanoderis* (Bally) Petch (1924)
- *Beauveria sulfurescens* (van Beyma) J.J. Taylor (1970)
- *Beauveria tenella* (Sacc.) MacLeod (1954)
- *Beauveria vexans* (Pettit) Petch (1926)

- *Beaveria delacroixii* (Sacc.) Petch (1924)
- *Botrytis bassiana* Bals.-Criv. (1835)
- *Botrytis bassiana* subsp. tenella (Sacc.) Sacc. (1882)
- *Botrytis bassiana* var. lunzinensis Szilvinyi (1941)
- *Botrytis delacroixii* (Sacc.) Delacr. (1893)
- *Botrytis effusa* Beauverie (1911)
- *Botrytis necans* Massee (1914)
- *Botrytis paradoxa* Bals.-Criv. (1835)
- *Botrytis stephanoderis* Bally (1923)
- *Botrytis tenella* Sacc. (1881)
- *Isaria citrinula* Speg. (1910)
- *Isaria densa* (Link) Giard (1891)
- *Isaria shiotae* Kuru (1931)
- *Isaria tenella* (Sacc.) Giard (1891)
- *Isaria vexans* Pettit (1895)
- *Penicillium bassianum* (Bals.-Criv.) Biourge, 1923
- *Penicillium delacroixii* (Sacc.) Biourge (1923)
- *Penicillium densum* (Link) Biourge (1923)
- *Spicaria bassiana* (Bals.-Criv.) Vuill. (1910)
- *Spicaria delacroixii* (Sacc.) Vuill. (1910)
- *Spicaria densa* (Link) Vuill. (1910)
- *Sporotrichum densum* Link (1809)
- *Sporotrichum globuliferum* Speg. (1880)
- *Sporotrichum larvatum* Peck, 1879
- *Sporotrichum larvicola* Peck (1887)
- *Sporotrichum minimum* Speg. (1881)

- *Sporotrichum sulfurescens* Van Beyma (1928)
- *Stachylidium bassianum* (Bals.-Criv.) Mont. (1856)
- *Trichoderma minimum* (Speg.) G. Müller (1965)
- *Tritirachium shiotae* (Kuru) Langeron (1947)

Local Common Names

- **Germany:** Parasit: Arthropoden

EPPO code

- **BEAUBA (*Beauveria bassiana*)**

Taxonomic Tree

- Domain: Eukaryota
- Kingdom: Fungi
- Phylum: Ascomycota
- Subphylum: Pezizomycotina
- Class: Sordariomycetes
- Subclass: Hypocreomycetidae
- Order: Hypocreales
- Family: Cordycipitaceae
- Genus: Beauveria
- Species: *Beauveria bassiana*

Discovery and name

The species is named after the Italian entomologist Agostino Bassi ،who discovered it in 1835 as the cause of the muscardine disease of domesticated silkworms. It was formerly also known as *Tritirachium shiotae.*

Morphology of the fungus

In culture ، *B. bassiana* grows as a white mold.On most common cultural media, it produces many dry, powdery conidia in distinctive white spore balls . Each spore ball is composed of a cluster of conidiogenous cells. The conidiogenous cells of *B. bassiana* are short and ovoid, and terminate in a narrow apical extension called a rachis.

The rachis elongates after each conidium is produced, resulting in a long zig-zag extension.The conidia are single-celled ،haploid ،and hydrophobic.

Classification of Entomopathogenic Fungi

Class: plant kingdom

Mastigomycotina

Class: Deuteromyceteae

Or : Moniliales

Fam : Moiliaceae

Beauveria & Metarhizium

White muscardine disease

The insect disease caused by the fungus is called white muscardine disease.When the microscopic spores of the fungus come into contact with the body of an insect host; they germinate, penetrate the cuticle ‹and grow inside, killing the insect within a matter of days.

Afterwards, a white mold emerges from the cadaver and produces new spores.A typical isolate of *B .bassiana* can attack a broad range of insects; various isolates differ in their host range.The factors responsible for host susceptibility are not known.

Beauveria bassiana parasitizing the Colorado potato beetle has been reported to be, in turn ‹the host of a mycoparasitic fungus *Syspastospora parasitica*.This organism also attacks related insect-pathogenic species of the Clavicipitaceae.

Fig. (4): *Beauveria bassiana*, Mass rearing on Rice medium[131]

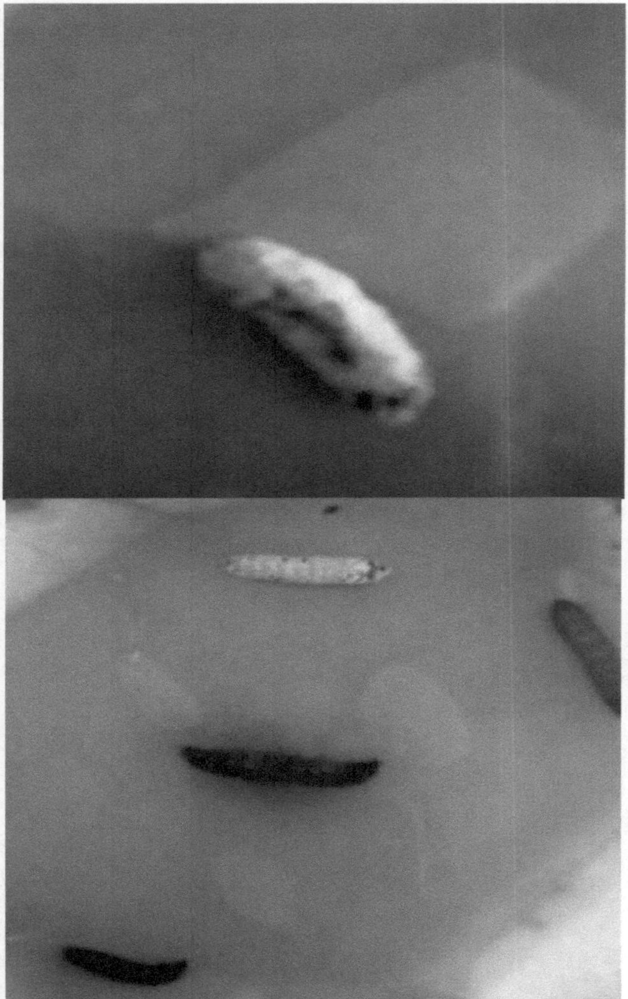

Fig. (5): ***Beauveria bassiana*****, infection insects,** [132, 133]

The life cycle of *B. bassiana*

B. bassiana has a dimorphic mode of growth. In the absence of the specific insect host *Beauveria* passes through an asexual vegetative life cycle that includes germination, filamentous growth and the formation of sympodulo conidia.

In the presence of its host insect, *Beauveria* switches to the pathogenic life cycle.

The conidiospores germinate on the surface of the cuticle and the germinated hyphal tubes penetrate the insect's integument directly. When having penetrated the cuticle, the fungus alters its growth morphology to a yeast-like phase and produces hyphal bodies, which circulate in the haemolymph and proliferate by budding.

Following the death of the host, fungal growth reverts back to the typical hyphal form (the saprotrophic stage). The ability to convert to the yeast-like phase may be a prerequisite for pathogenicity.

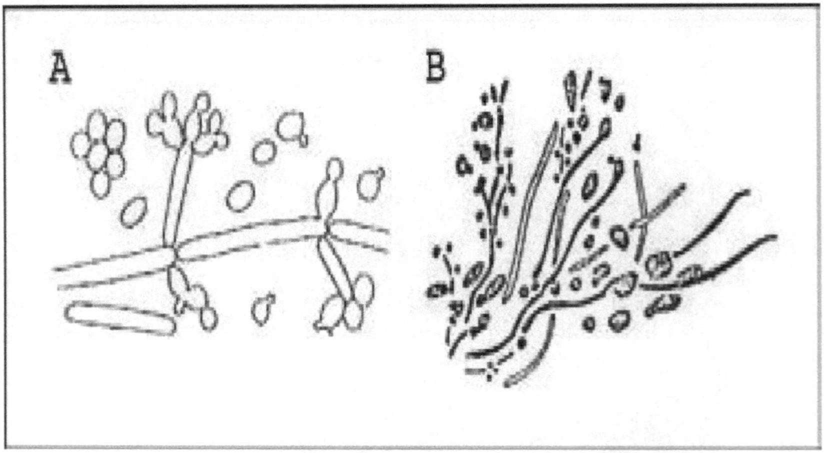

Fig. (6): Dimorphic growth mode of *B. bassiana*. (A) Yeast. Like parasitic phase when infecting susceptible species. (B) Saprobic phase shows filamentous hypha.

Counting spore number

The spore suspension was throughly vortexed and diluted with sterile 0.05% (v/v) Tween 80. The spore number of the desired dilution was counted with a Hemacytometer (Superior, Marienfeeld, Germany) and a light microscope (Olympus BH2, Japan) at a magnification of × 400.

The infection process

In contrast to bacteria and viruses that pass through the gut wall from contaminated food, fungi have an unique mode of infection. They reach the hamocoel through the cuticle or possibly through the mouth parts.

Ingested fungal spores do not germinate in the gut and are voided in the faeces.

The death of the insect results from a combination of factors: mechanical damage resulting from tissue invasion, depletion of nutrient resources and toxicosis, [30].

Adhesion and germination of conidia

Attachment of a fungal spore to the cuticle surface of a susceptible host represents the initial event in the establishment of mycosis. For most entomopathogenic fungi host location is a random event and attachment is a passive process with the aid of wind or water.

It was found that dry spores of *B. bassiana* possess an outer layer composed of interwoven fascicles of hydrophobic rodlets. This rodlet layer appears to be unique to the conidial stage and has not been detected on the vegetative cells.

The adhesion of dry spores to the cuticle was suggested to be due to non-specific hydrophobic forces exerted by the rodlets [31].

In addition, lectins, a kind of carbohydrate binding glycoproteins, have been detected on the conidial surface of *B. bassiana*.

It was also suggested that lectins could be involved in binding between conidia and the insect cuticle. The exact mechanisms responsible for the interaction between fungal spores and the cuticle remain to be determined [32].

After the pathogen reaches and adheres to the host surface, it proceeds with rapid germination and growth which are profoundly influenced by the availability of nutrients, oxygen, water, as well as pH, and temperature, and by the effects of toxic host-surface compound.

Generally, fungi with a broad host range germinate in culture in response to a wide range of nonspecific carbon and nitrogen sources.

Entomopathogenic fungi with restricted host range appear to have more specific requirements for germination [33].

Formation of an infection structure

Entomopathogenic fungi invade their hosts by direct penetration of the host cuticle. The cuticle has two layers, the outer epicuticle and the procuticle.

The epicuticle is a very complex thin structure that lacks chitin but contains phenol-stabilized proteins and is covered by a waxy layer containing fatty acids, lipids and sterols [34].

The procuticle forms the majority of the cuticle and contains chitin fibrils embedded into a protein matrix together with lipids and quinones [35].

Protein may account for up to 70% of the cuticle. In many areas of the cuticle the chitin is organized helically giving rise to a laminate structure.

In common with many entomopathogenic fungi, *B. bassiana* conidia germinate on the host surface and differentiate an infection structure termed appressorium.

Penetration of the cuticle

Pathogenic fungi need to penetrate through the cuticle into the insect body to obtain nutrients for their growth and reproduction. Entry into the host involves both enzymic degradation and mechanical pressure as evidenced by the physical separation of lamellae by penetrated hyphae.

A range of extracellular enzymes that can degrade the major components of insect cuticle, including chitinases, lipases, esterases and at least four different classes of proteases, have been suggested to function during the fungi pathogenesis.

The production of cuticle-degrading enzymes by *M. anisopliae* during infection structure formation on *Calliphor vomitoria* and *Manduca sexta* has been investigated by biochemical and histochemical analysis both *in vivo* and *in vitro*.

Among the first enzymes produced on the cuticle are endoproteases (termed PR1 and PR2) and aminopeptidases, coincident with the formation of appressoria. N-Acetylglucosaminidase is produced at a slow rate as compared to the proteolytic enzymes.

Chitinase and lipase activities were not detected [36].

Although the complex structure of the insect cuticle suggests that penetration would require the synergistic.

Production of toxins

There is considerable circumstantial evidence from deuteromycete pathogens for the involvement of fungal toxins in host death. The action of cytotoxins is suggested by cellular disruption prior to hyphae penetration. Behavioural symptoms such as partial or general paralysis, sluggishness and decreased irritability in mycosed insects are consistent with the action of neuromuscular toxins [142].

B. bassiana and *M. anisopliae* produced significant amounts of toxic compounds within their hosts.

For example, the toxins Beauvericin, Beauverolides, Bassianolide and Isarolides

Host defense systems

In order to prevent invasion by fungi, insects have evolved various defense mechanisms. The defensive arsenal of insects contains both passive structural barries, such as the cuticle, and a cascade of active responses to pathogens that gain access to the hemocoel.

These active responses include melanization, cellular reactions, humoral reaction to recognize the non-self pathogen, and production of protease inhibitors.

Melanization:

the oxidation of phenolic compounds to dihydroxyphenylalanine, typified by the production of brown or black melanic pigments, is a common feature of the response of many insects to fungal infection.

Melanin may partially shield cuticle from enzymatic attack or may be toxic to fungi. However, such protection is incomplete. The investigation from [32] indicates that melanization is primarily an effective defence against weak or slow growing pathogens, but is ineffective against more virulent fungi.

Cellular reactions

Once the cuticle and epidermis have been breached the invading fungus is faced with the defence systems of the hemolymph.

The responses to mycopathogens within the haemocoel include phagocytosis, encapsulation and nodulation. However, the effect on fungal elements is uncertain. With the arbitrary injection method, [143] found that haemocytes of the migratory grasshopper, *M. sanguinipes*, encapsulate viable conidia of *B. bassiana*, however they fail to suppress conidial germination within the nodule.

It was suggested that the production of toxins and extracellular proteases by *B. bassiana* could trigger the evadation of encapsulation.

Production of protease inhibitors:

Host-produced protease inhibitors, which inhibit cuticle-degrading enzyme activities of pathogens, may contribute to insect defence systems. Such compounds have been isolated from the serum of *Anticarsia gemmatalis* larvae which were resistant to infection by *Nomuraea rileyi* [38].

Aspects of using entomopathogenic fungi as bio-control agents

The advantages of using fungi as insecticides are:

(1) Their high degree of specificity for pest control. Fungi can be used to control harmful insect pests without affecting beneficial insect predators and nonharmful parasites.

(2) The absence of effects on mammals and thus the reduction of the hazards normally encountered with insecticide applications, such as pollution of the environment.

(3) The lack of problems caused to insect resistance and prolonged pest control.

(4) A high potential for further development by biotechnological research.

(5) High persistence in the environment provide long-term effects of entomopathogenic fungi on pest suppression.

However, there are also a number of constraints on the use of fungi as insecticides:

(1) 2-3 weeks are required to kill the insects whereas chemical insecticides may need only 2-3 hours.

(2) Application needs to coincide with high relative humidity, low pest numbers and a fungicide free period.

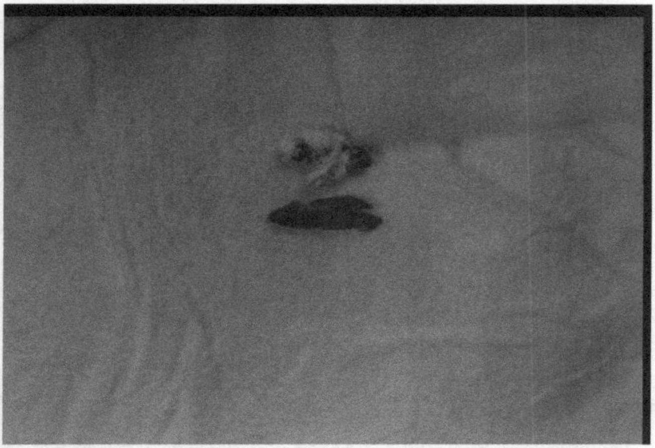

Fig. (7): Pupa of *S. ocellatella* infected with *B. bassiana* (Abdel-Raheem, 2005)

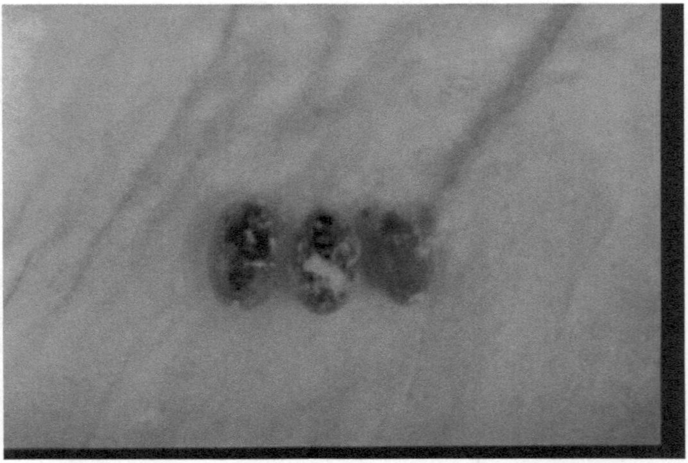

Fig. (8): Adult of *Cassida vittata* infected with. *B. bassiana* (Abdel-Raheem, 2005)

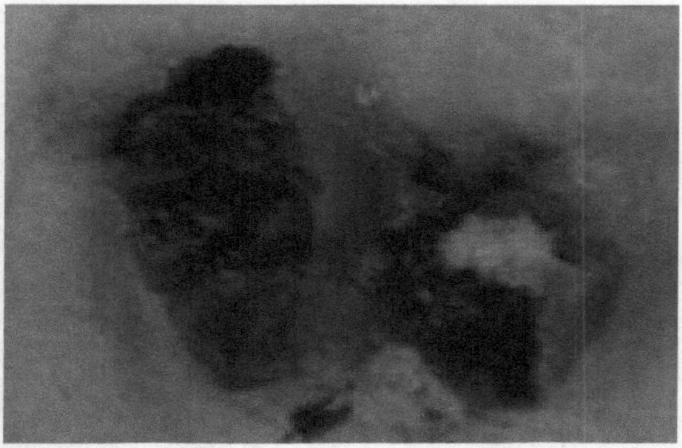

Fig. (9): Pupa of *C. vittata* infected with *B. bassiana* (Abdel-Raheem, 2005)

Fig. (10): ***Rhynchophorus ferrugineus*** **(red palm weevil); natural enemy. Dorsal and ventral views of** ***Beauveria bassiana*****, sporulating on the cadaver of a red palm weevil.**

Genetic engineering of entomopathogenic fungi

A more widespread use of fungi for bio-control will depend on improvements of wild-type strains by combining characteristics of different strains and mutants. Two type of improvements may be considered: (i) improvement the efficacy of the insecticide, e.g., by reducing the dose necessary to kill the insects, by reducing the time to kill the pest or decreasing crop damage caused by the pest by reducing the feeding time; (ii) expanding the host range.

Essential for the development of a hypervirulent strain is a complete understanding of the remarkable pathology of fungi infections.

Molecular biology provides the necessary tools for dissecting the mechanisms of pathogenesis and in the longer term for producing recombinant organisms with new characteristics.

Initial development towards these goals has occurred with *M. anisophliae* and to a much lesser extent with *B. bassiana* [39].

Genetic transformation systems, which are an essential part of modern fungal research, and is necessary for the experimental manipulation of virulence genes *in vitro* and *in vivo,* have been established [40].

The success of utilizing these procedures depends on the availability of selectable transformation markers.

Transformation techniques have been used to isolate specific pathogenic genes, investigate virulence determinants of *M. anisopliae*, and to produce a strain with enhanced virulence.

Unravelling the molecular mechanisms of fungal pathogenesis in insects will provide the basis for the genetic engineering of entomopathogenic fungi.

Use in biological control of insects

Beauveria bassiana can be used as a biological insecticide to control a number of pests such as termites ،whiteflies ،and many other insects .Its use in the control of malaria-transmitting mosquitos is under investigation.

As an insecticide, the spores are sprayed on affected crops as an emulsified suspension or wettable powder or applied to mosquito nets as a mosquito control agent.

Beauveria bassiana parasitizes a very wide range of arthropod hosts, and so should be considered a nonselective biological insecticide. It should not be applied to flowers visited by pollinating insects

Known targets include

Aphids - Whiteflies -Mealybugs -Psyllids -Lygus bugs - Chinch bug - Grasshoppers -Thrips-Termites -Fire ants-Flies-Stem Borers -Fungal gnats - Shoreflies -Beetles

Coffee borer beetle-Colorado potato beetle-Mexican bean beetle-Japanese beetle -Boll weevil -Cereal leaf beetle -Bark beetles -Black vine weevil - Strawberry root weevil- Caterpillars- European corn borer - Codling moth- Douglas fir tussock moth– Silkworm -Mites.

The fungus rarely infects humans or other animals, so it is generally considered safe as an insecticide. However, at least one case of human infection by *B .bassiana* has been reported in a person with a suppressed immune system.

Additionally, like any powder, the spores may exacerbate breathing difficulties.Wagner and Lewis reported the ability of *B. bassiana* to grow as an endophyte in corn.

The spore producing apparatus in *Beauveria* is small and delicate. It is very difficult to make a good slide mount to show the fine details of origin and development (= ontogeny) of the spores.

Photomicrograph above shows three conidiogenous cells (= spore mother cells) on a hypha (= fungal filament) of *Beauveria bassiana*. The conidia (= asexual spores)) are produced in acropetal succession (= youngest at the tip) on an elongating conidiogenous cell (= cell that produces conidia).

Conidium is the name given to the non sexual spore that germinates by means of a germ tube that is produced in the Hyphomycetes (= class) of the Deuteromycota (= Division for asexual fungi).

Sympodial Spore Production in Beauveria

In *B. bassiana* the spores are produced sympodially. A spore is produced at the tip of the mother cell and the growth of the mother cell ceases.

A new growing point initiates just below this terminal spore, grows past it, and a second spore is produced at a higher level.

This uses up the new growing point and a third growing point is then initiated just below the second spore.

Every time a spore is produced the hyphal tip is used up and a new growing point is produced.

In this way a succession of spores is produced with the youngest spore at the tip (= i.e acropetal succession) and the spore head gets longer and longer.

When all the spores secede (= are dislodged) the spore-bearing tip of the conidiogenous cell has a zig zag appearance and is referred to as a rachis (this term is also used for the seed head in wheat when all the seeds are gone).

Biology of Beauveria

Beauveria bassiana is an aggressive parasite of many different insect host species. Not only does it have a wide host range but insects are attacked at larval or adult stages.

The spores are tiny, measuring only a few microns. The hyphae (hyaline) and so and spores are non pigmented colonies appear white in cultures or tufts of white mycelium bearing masses of powdery spores burst out through the body parts of infected insects as in the cicada, hawk moth and chinch bugs.

For something really dramatic look at the synnematal form of *Beauveria* I found growing on a hawk moth. And right next to that is a couple of hairy chinch bugs that I recovered from the chinch bug population inhabiting my lawn.

Beauveria bassiana strain 447 is a naturally occurring fungus that is approved for indoor use as an insecticide in bait stations to control individual fire ants and other ants (but not colonies.)

For something really dramatic look at the synnematal form of *Beauveria* I found growing on a hawk moth. And right next to that is a couple of hairy chinch bugs that I recovered from the chinch bug population inhabiting my lawn.

- The fungal spores infect the ants ،which soon die.
- The fungus is not toxic or infective to mammals, and exposure to the public and the environment will be minimal to non-existent.
- Therefore, no adverse effects are expected on children, adults, pets, or the environment when the bait stations are used as directed.

Description of the Active Ingredient

- *Beauveria bassiana* strain 447 belongs to a group of fungi that are found throughout the world.
- The infection begins when a fungal spore starts growing on the outer skeleton of a host insect, and the infected insect soon dies.

Assessing Risks to Human Health

- Whether or not a substance poses a risk to humans or other organisms depends on two factors:
- How toxic the substance is, and how much of it an organism is exposed to.
- Therefore, the EPA considers toxicity data and exposure data in determining whether to approve a pesticide for use.

No adverse human health effects are expected from use of *Beauveria bassiana* strain 447 in bait stations.

- The bait stations are designed to prevent adults, children, or pets from being exposed to the fungus.
- In addition, laboratory studies show that the active ingredient is not toxic or infective to laboratory animals when eaten or inhaled.
- The fungus grows only at temperatures lower than human body temperature, so no infection is expected.

Assessing Risks to the Environment

- No harmful environmental effects are expected because the bait stations are approved only for use indoors.
- Because the active ingredient is firmly attached to the inside of the bait station, no outdoor exposure is expected.
- EPA will require additional studies if the bait stations are proposed for outdoor or other additional uses.

Products Directed Against Public Health Pests

- EPA defines a public health pest as any organism that can cause or transmit human disease, or can cause human discomfort or injury.
- Fire ants are categorized as a public health pest.
- To help protect the public's health, EPA requires registrants of products used against public health pests to demonstrate that the product meets specific standards for effectiveness as well as for safety.

At present improved fungal mass production and formulations are commercially available.

Modern techniques in genetic engineering and biotechnology are extremely helpful in manipulating the desired traits of entomopathogenic fungi, which can further improve its bioactivity˙

Entomopathogenic fungi were among the first organisms to be used for the biological control of pests. More than 700 species of fungi from around 90 genera are pathogenic to insects.

Most are found within the deuteromycetes and entomophthorales. some insect-pathogenic fungi have restricted host ranges, for example, *Aschersonia aleyrodis* infects only scale insects and whiteflies, while other fungal species have a wide host range, with individual isolates being more specific, for example, *Metarhizium anisopliae* and *Beauveria bassiana*.

The commercial produced entomopathogenic fungi and their targeted hosts are shown in Table 4.

Fungal species such as *M. anisopliae* and *B. bassiana* are well characterized in respect to pathogenicity to several insects and they have been used as agents for the biological control of agriculture pests worldwide.

In Colombia, about 11 companies offer at least 16 products based on the entomopathogenic fungi *B. bassiana*. These products are used not only in the coffee crop but also in other crops such as cabbage, corn, bean, tomato, potato. They are also used to treat public disease vectors (e.g., flies and mosquitoes) [41].

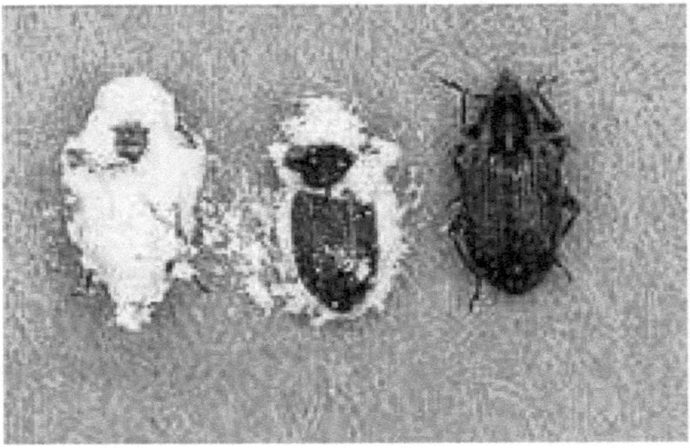

Fig. (11): Biological control of insect pest (alfalfa weevil). Left and Center, Adult insect killed by *Beauveria bassiana*. Right, Healthy adult insect (control).

Table (4): Entomopathogenic fungi in commercial and experimental production (adapted from Khachatourians, 1986)

Fungus	Targets	Product
Beauveria bassiana	Colorado potato beetle Coding moth European corn borer Pine caterpillar	Boverin Mycotrol®
Culicinomyces clavisporus	Mosquito larvae	
Hirsutella thompsonii	Citrus rust mite	Mycar
Metarhizium anisopliae	Spittle bug Sugarcane frog hopper	Metaquino Meta-Sin®
Verticillium lecanii	Aphids Coffee green bug Greenhouse whitefly Thrips	Vertalec

Metarhizium species

- *Metarhizium* species is being tested as natural enemy of many soil insect pests.
- *Metarhizium* can be an important control agent of aphids.
- In Indonesia (West Sumatra), *Metarhizium* is used as a spraying solution for control of different pest insects in chili.
- In Philippines, *Metarhizium* effectively reduced populations of rhinoceros beetles in coconut [45].
- In the United States *Metarhizium* is commercially available as a foliar spray.

Metarhizium is usually called the green fungus of insects. The fungus has white mycelia within the body, but when it is ready to form spores, the spores coat the killed host with a velvety covering of olive-green.

Fig.(12): Green June beetle grub skilled by *Metarhizium*.

Fig.(13): Japanese beetle grub infected with *Metarhizium*.

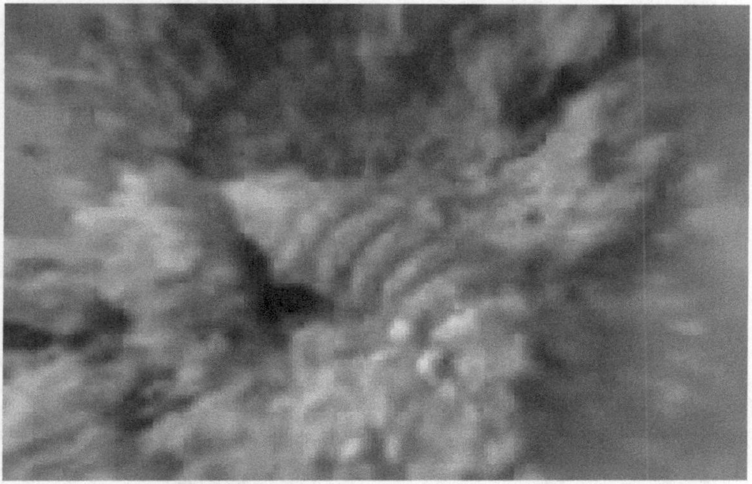

Fig.(14): ***Agrotis ipsilon*** **with** ***Metarhizium*** **[46]**

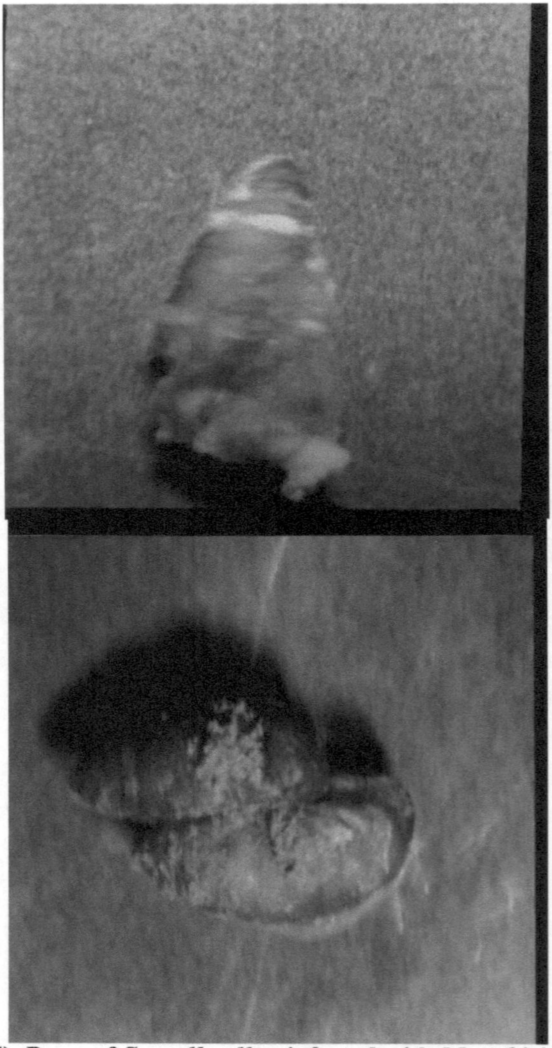

Fig.(15): Pupa of *S. ocellatella* infected with *Metarhizium* [47]

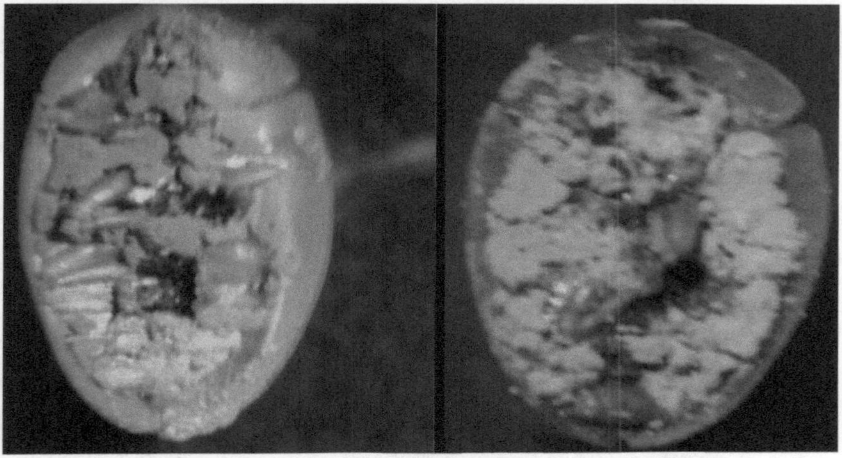

Fig.(16): Adult of *C. vittata* infected with *Metarhizium* [47]

Fig.(17): Rice grains infected with *Metarhizium* [48]

Fig.(18): Rice grains infected with *Metarhizium* [48]

Mode of action and symptoms

- Fungi penetrate the skin of insects.
- Once inside the insect, the fungus rapidly grows throughout the body.
- Death is caused by injury to the tissue or, occasionally, by toxins produced by the fungus.
- The fungus grows out through the insect's skin to produce spores that can sometimes be seen as "powder" or "cotton" on the skin of the dead insect.
- When spread by wind, rain or contact with other insects, the spores can spread the infection.
- Infected insects stop feeding and become lethargic.
- They may rapidly die, sometimes in a position still attached to a leaf or stem.
- The dead insect's body may be firm or it may be an empty shell.
- The fungus is often seen as "hairs" or "cotton" in various colors on parts of the insects body.

Effectiveness

- Insect-pathogenic fungi usually need high humidity to cause infection.
- Natural infections are therefore most common during the wet season.
- The effectiveness of fungi against insect pests depends on many factors:

Having the correct fungal species with the susceptible insect life stage, at the appropriate humidity, soil texture (if you are trying to reach ground-dwelling insects), and temperature.

The fungal spores, which can be carried by wind or water, must contact the pest insect to cause infection.

- When insect-pathogenic fungi are applied, for example through a spore-solution, good spray coverage of the plants is essential.

How to protect and help fungi

- Many insect-pathogenic fungi live in the soil.
- There is evidence that application of some soil insecticides, fungicides and herbicides can inhibit or kill these fungi.
- For example, even very low concentrations of some herbicides can severely limit the germination and growth of *Beauveria bassiana* fungal spores in soil samples.

Primary or secondary

- When you leave a plate of food with for example some chicken meat in your kitchen and leave it for a few days, you may find the meat covered with fungus when you look again.

- This fungus is called a *secondary infection*:
- It was not the reason the chicken died (the chicken probably died because a person killed it), and the fungus started growing after the chicken was dead.

Similarly, when insects are dead, some fungi may start growing on the dead insect.

- This is also called a secondary infection.
- These fungi that cause secondary infection are part of the "trash men team" of nature:
- They make sure dead things are decomposed quickly.
- When the fungi actually cause a living insect to die, like insect-pathogenic fungi do,
- It is called a *primary infection*

Interesting, but who cares?

- When searching for beneficial fungi in the field, it is important to distinguish between fungi causing primary infection (pathogens) and those causing a secondary infection.
- Fungi that cause secondary infections are not really pathogens (will not infect and Kill healthy insects).

Application of Entomopathogenic Fungi for Insect Pests Control

Bioassay of isolated fungi against Many Insects:

Larvae of, *G. mellonella*

G. mellonella is usually used as a test insect for assessment of virulence of entomopathogenic microorganisms including fungi, [49]. Fungal concentrations around $1x10^7$ spores/ml are frequently used in these bioassayes. Target insect may be directly used if available. Gupta, [49], compared the virulence of five strains of *B. bassiana* against the larvae of *G. mellonella.*

% of mortality among *G. mellonella* larvae exposed for 10 days to serial concentrations were ((C_1) $1x10^3$, (C_2) $1x10^4$ and (C_3) $1x10^5$ spores /ml.). *B. bassiana* the most virulent fungus to *G. mellonella* larvae followed by *M. anisopliae* then *V. lecanii.* At the lowest concentration of (1x 10^3 spores/ ml) mortality of *G. mellonella* larvae ranged 14.4 to 99.2 %, 10.4 to 90.4% and 10.4 to 84% due to *B. bassiana, M. anisopliae* and *V. lecanii,* respectively. At the highest concentration, the three fungi reached their maximum effect at 10 days after application being 100%, 98.4% and 97.6%, respectively, [26]. Virulence depended greatly on the concentration and the exposure time.

B. bassiana the fastest effect against larvae of *G.mellonella*, followed by *M. anisopliae* then *V. lacanii.* Strong relationship was found between either the concentration of the fungus or the time of exposure and the resulted mortality in larvae of *G. mellonella*. Values of squared correlation coefficient (R^2) were always over 0.9, [26].

Using of *B. bassiana* and *M. anisopliae* were the most effect to decrease the total number of *C. vittata* (Vill.) The fungus *B. bassiana* was the most effect to controlling the tortoise beetle, *C. vittata* than the Fungus *M. anisopliae.*

Table (5) the susceptibility of different stages of tortoise beetle *C. vittata* to entomopathogenic fungus *B. bassiana at* different concentrations, [12]:

Treated stages	%Mortality at indicated concentrations					LC 50	Fudicial limits 95 %	Slope ± SE
	10^6	$5x10^6$	10^7	$5x10^7$	10^8			
1st instar larvae	50	75	85	95	100	1.08×10^6	($6.35x10^5$ -1.6×10^6)	1.09± 0.121.
2nd instar larvae	40	55	80	90	100	2.24×10^6	($2.92x10^5$ –$5.42x10^6$)	1.1 ± 0.2
3rd instar larvae	40	50	65	90	100	$1.49x 10^6$		0.79± 0.31
4th instar larvae	80	80	85	80	100	$1.03x10^5$	($0.2x10^5$ –$602x10^5$)	2.33 ±0.73
5th instar larvae	50	60	70	70	85	$1.04x 10^6$	($2.16 x10^5$–$2.38x10^6$)	0.43 ± 0.08
Pupa	50	65	65	75	95	$1.32x10^6$	($0.013x10^6$–$6.04x10^6$)	0.59 ±0.17
Adult	80	85	85	95	100	$0.4x10^5$	($0.005x10^5$ –$3.13 x10^5$)	2.55 ± 0.76

Table (6) the susceptibility of different stages of tortoise beetle *C.vittata* to entomopathogenic fungi *M. anisopliae at* different concentrations, [12]:

Treated Stages	%Mortality at indicated concentration					LC $_{50}$	Fudicial limits 95 %	Slope ±SE
	10^{6}	5x 10^{6}	10^{7}	5x10^{7}	10^{8}			
1st instar larva	60	70	70	95	100	3.68 x10^{6}		1.4± 0.5
2nd instar larva	60	55	55	60	75	2.45x10^{5}	(6.86x10^{5}- 1.2x 10^{6})	0.16 ± 0.08
3rd instar larva	60	65	70	85	100	8.49 x10^{5}		0.69 ± 0.22
4th instar larva	60	60	60	70	100	9.8 x10^{5}		0.52 ± 0.49
5th instar larva	55	55	65	75	90	1.22 x10^{6}	(0.03x10^{6} - 5.79x10^{6})	0.5 ± 0.14
Pupa	45	50	55	60	70	3.65 x10^{6}	(6.32 x10^{5} – 8.72x10^{6})	0.3 ± 0.08
Adult	60	75	70	95	100	2.27 x10^{5}	(0.9x10^{5} – 1.12 x10^{6})	2.15 ± 0.76

Table (7) the susceptibility of different stages of sugar beet mining moth *S. ocellatella* to entomopathogenic fungus *B. bassiana at* different concentrations, [12]:

Treated stages	%Mortality at indicated concentrations					LC_{50}	Fudicial limits 95 %	Slope ± SE
	10^6	$5x10^6$	10^7	$5x10^7$	10^8			
4th instar larva	60	75	60	60	70	$0.4 x10^5$		0.04 ± 0.14
Pupa	75	90	95	100	100	$2.99 x10^5$	($0.9 x10^5$ –5.63 x 10^5)	1.2 ± 0.19

Table (8) the susceptibility of different stages of sugar beet mining moth *S. ocellatella* to entomopathogenic fungi *M. anisopliae at* different concentrations, [12]

Treated stages	%Mortality at indicated concentrations					LC $_{50}$	Fudicial limits 95 %	Slope ± SE
	10^6	$5x10^6$	10^7	$5x10^7$	10^8			
4th instar Larva	70	70	75	90	95	$2.27x10^5$	($6.14 x10^5$ – $6.53 x10^5$)	0.52 ± 0.09
Pupa	20	20	35	70	75	$1.92 x10^7$	(6.14 x10 6 –$9.6 x10^7$)	0.88 ± 0.18

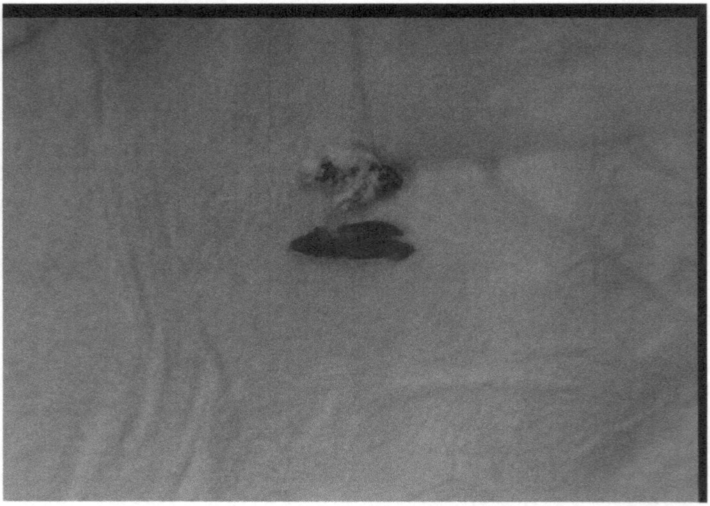

Fig . (19): Pupa of *S. ocellatella* infected with *B. bassiana*, [12]

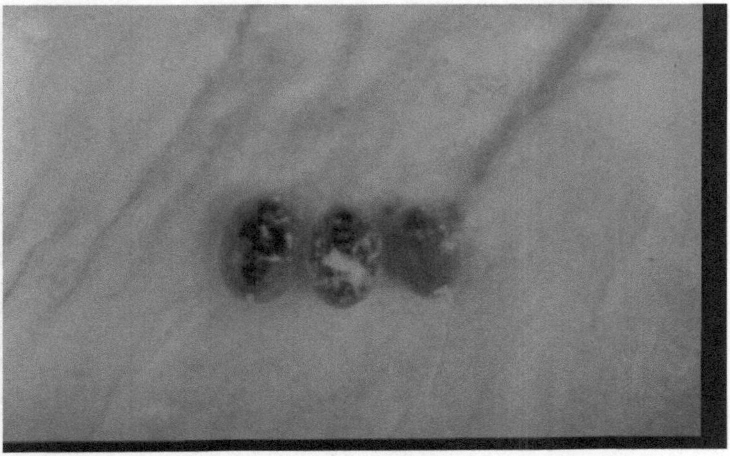

Fig. (20): Adult of *C. vittata* infected with. *B. bassiana*, [12]

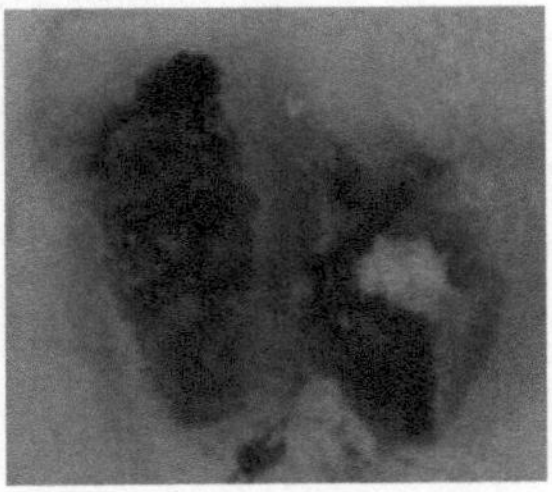

Fig. (21): Pupa of *C. vittata* infected with *B. bassiana*, [12]

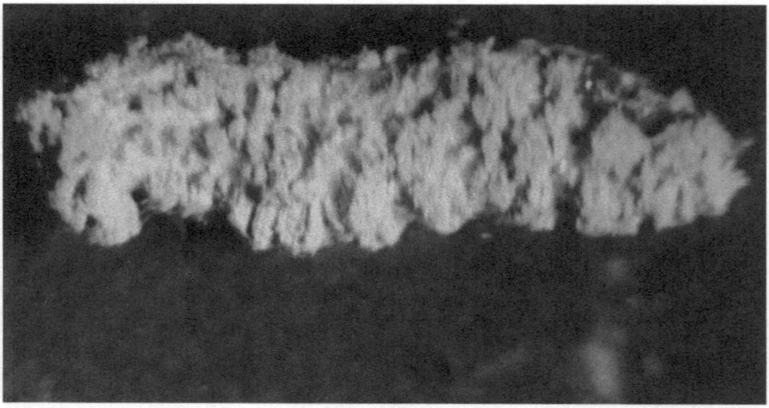

Fig. (22): Larvae of *S. ocellatella* infected with *M. anisopliae*, [12]

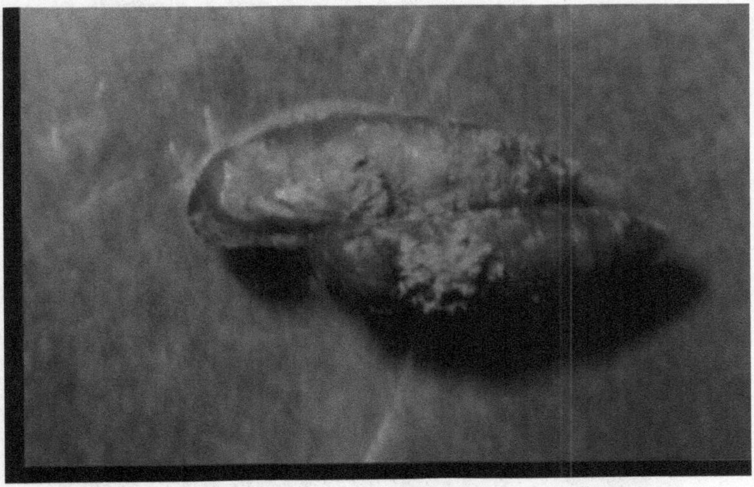

Fig . (23): Pupa of *S. ocellatella* infected with *M. anisopliae*, [12]

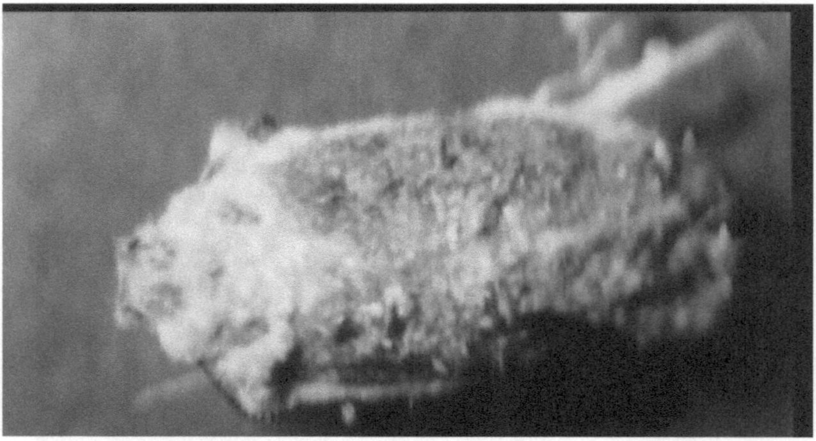

Fig. (24): Adult of *S. ocellatella* infected with M. *anisopliae*, [12]

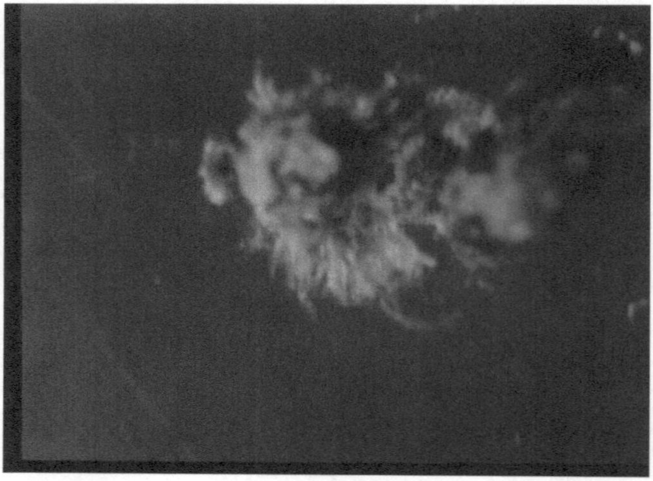

Fig. (25): 1st instar larva of *C. Vittata* infected with *M. anisopliae*, [12]

Fig. (26): 2nd instar larva of *C. vittata* infected with*M. anisopliae*, [12]

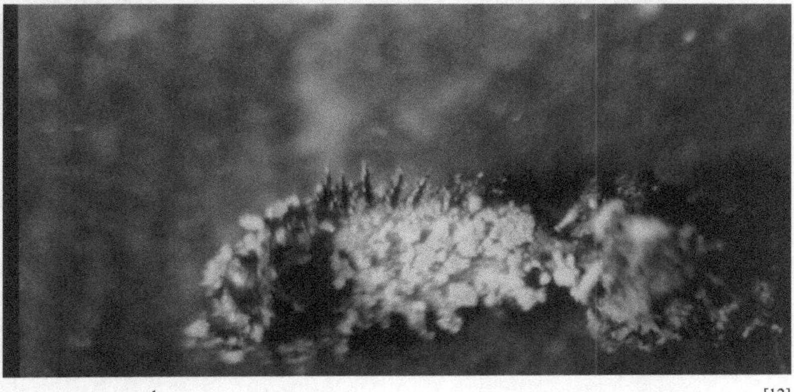

Fig. (27) : 3rd instar larva of *C. vittata* infected with *M. anisopliae*, [12]

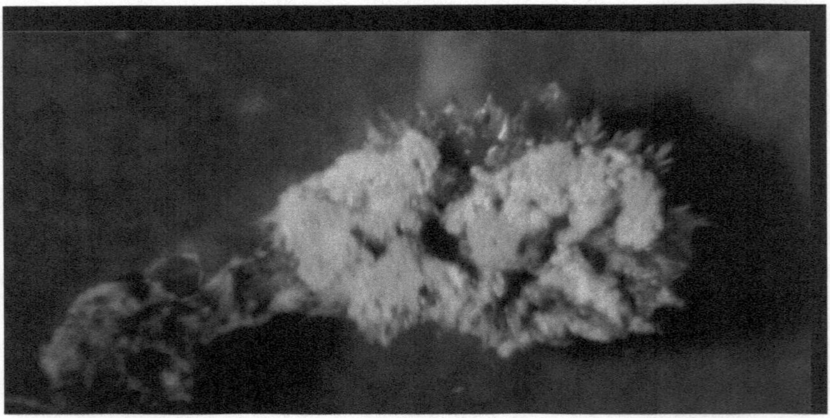

Fig. (28) : 4th instar larva of *C. vittata* infected with *M. anisopliae*, [12]

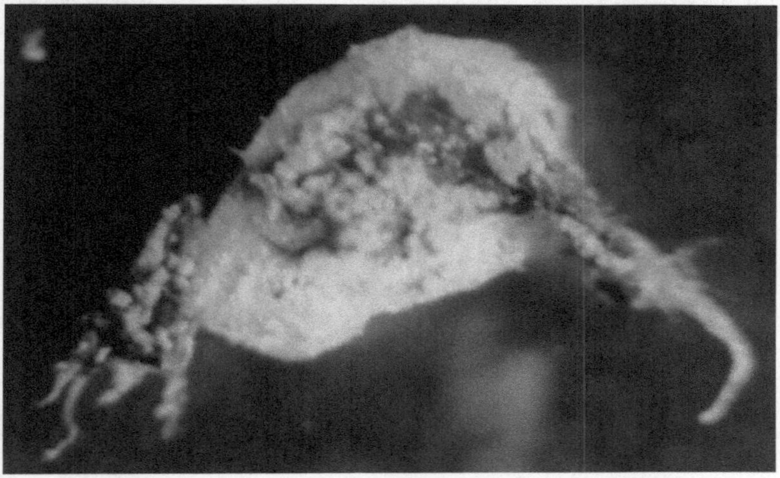

Fig. (29): 5th instar larva of *C. vittata* infected with *M. anisopliae*, [12]

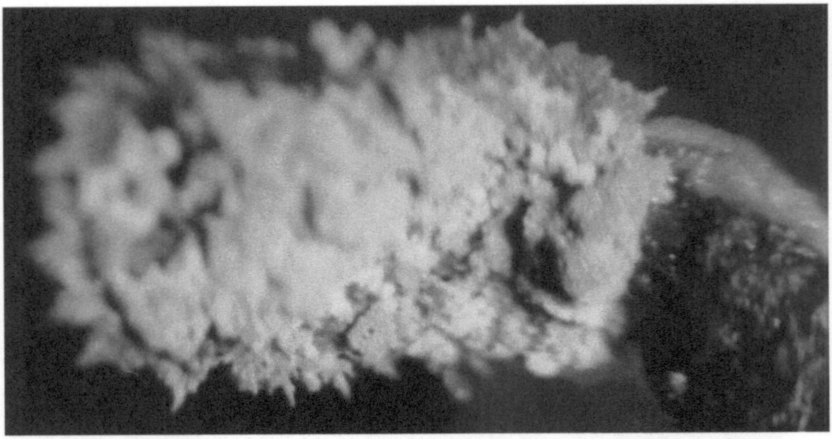

Fig. (30): Pre-pupa of C. *vittata* infected with *M. anisopliae*, [12]

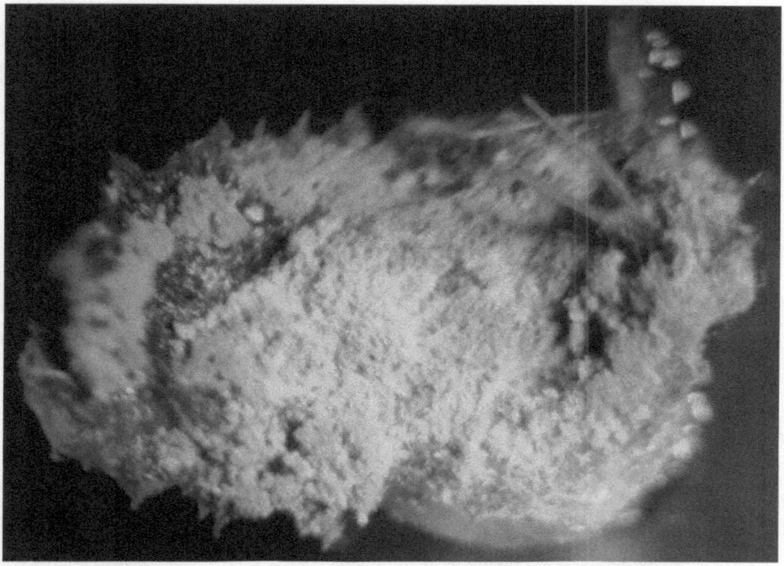

Fig. (31) Pupae of C. *vittata* infected with *M. anisopliae*, [12]

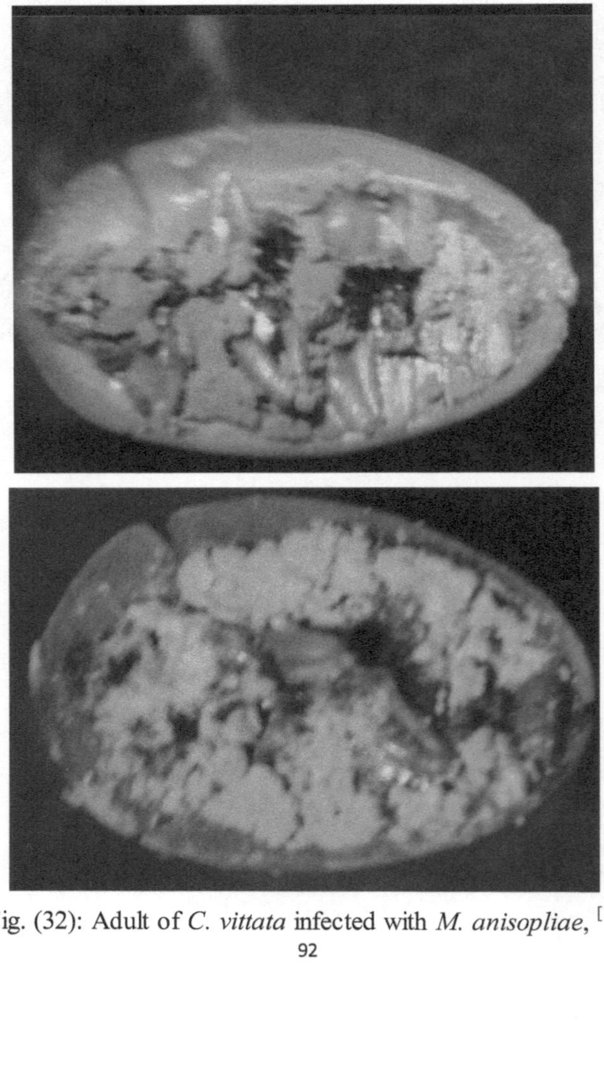

Fig. (32): Adult of *C. vittata* infected with *M. anisopliae*, [12]

Tomato leaf miner, *Tuta absoluta* (Meyrick)

Efficacy of 3 concentrations of *B. bassiana, M. anisopliae* and *V. lecanii* were prepared with concentrations of ($1x10^5$; $1x10^6$; $1x10^7$) were tested on *T. absoluta* larvae (Neonate "newly hatched", 2nd & 3rd instar), [51].

Eggs of *T. absoluta* were exposure to *B. bassiana, M. anisopliae* and *V. lecanii* to evaluate their effect on hatchability. The estimated LC_{50} of *V. lecanii* were 3.25 x 10^5 spores/ml, 5.47 x 10^5& 3.28 x $10^5$for neonate, 2nd instar & 3rd instar *T. absoluta* larvae, respectively. While the LC_{50} values of *B. bassiana* & *M. anisopliae* were (0.28 x 10^5 & 0.11 x 10^5), (0.45 x 10^5 & 0.46 x 10^5) and (0.32 x 10^5 & 0.27 x 10^5 conidia/ml) for neonate, 2nd instar & 3rd instar *T. absoluta* larvae, respectively. According to LC_{50} values, *B. bassiana* and *M. anisopliae* were most effective on larval phase of *T. absoluta* than *V. lecanii*, [51].

Daily mortality (%) of larval phase of *T. absoluta* fed in the newly hatched 2nd instar and 3rd instar larva with leaves treated with *V. lecanii*. The results revealed that when neonate larvae fed on *V. lecanii* the pathogen effect was evident by the third day of evaluation after exposure in the three concentrations (10^5; 10^6; 10^7spores/ml) with recorded mortality (13.3, 13.3, 33.3%) respectively, [52].

Thereafter, the values of the corrected mortalities of neonate larvae increased gradually from the 4th day after exposure until the last day (10th) with mortality 88.8% for the first concentration 10^5 spores/ml. For the second concentration (10^6 spores/ml) the mortality values were increased and reached its maximum in the 8th day of exposure (100% reduction). The mortality values reached its maximum in the 7th day and in the 5th day for the second and third concentration (10^6 & 10^7 spores/ml) and recorded 100% reduction for the two concentrations, [51].

The higher effective concentration of *V. lecanii* on neonate larvae of *T. absoluta* was 10^7spores/ml followed by 10^6 spores/ml while the other concentrations (10^5 spores/ml) showed moderate effect, [51].

For the second instar larva of *T. absoluta* the pathogen effect was evident by the fifth day for (10^5 spores/ml), by the 4th day for (10^6 spores/ml) and by the 3rd day for (10^7 spores/ml) of evaluation after exposure with corresponding mortalities (53.3, 40.0, and 6.6 %) for the three concentrations, respectively, [52]. The values of the corrected mortalities of 2nd instar larvae increased gradually until the 9th day after exposure to record 88.8% mortality for the first concentration (10^5 spores/ml), [51].

The mortality values reached its maximum in the 9th day, 8th day and in the 7th day for the three concentrations and recorded 100% reduction. Daily mortality (%) of larval phase of *T. absoluta* fed in the newly hatched "neonate", 2nd instar and 3rd instar larva with leaves treated with *B.bassiana*.

The higher effective concentration of *B. bassiana* on the 2nd instar larvae of *T. absoluta* resulted from the two concentrations 10^7 and 10^6conidia/ml evenly, followed by 10^7 and 10^6 conidia/ml, respectively, [51].

When 3rd instar larvae fed on *B. bassiana* the pathogen effect was evident by the 3rd day of evaluation after exposure in the first concentration (10^5 conidia/ml) with recorded mortality (20.0%).

While the pathogen effect was evident by the 2nd day of evaluation after exposure in the 1st, 2rd & 3th concentrations (10^5, 10^6 & 10^7) with recorded mortalities (6.7%) for the three concentrations, [51].

The potato tuber moth *Phthorimaea operculella* (ZELLER)

The LC_{50} of the target pest $1X10^6$ and 8.7 $X10^5$ spores /ml after *P. operculella* treated with *B. brongniartii* and *N. rileyi*, respectively. Under semi-field conditions, the corresponding LC_{50} recorded $1.20X10^6$ and 9.7 $X10^5$ spores /ml respectively), [52].

The effect of the entomopathogenic fungi against *P. operculella*, the number of eggs laid/ female recorded significant reduction to 42±1.7 and 33±8.9, after being treated with *B. brongniartii* and *N. rileyi* as compared to the control, respectively. The percentage of emerged adults was significantly decreased by (4%) as compared with the control, [52].

Cabbage Worm, *Pieris rapae* L.

Three isolates of *B. bassiana*, *M. anisopliae* and *V. lecanii* were evaluated against cabbage worm, *P. rapae* L. under laboratory and field conditions. *B. bassiana*, *M. anisopliae* and *V. lecanii* have not effect on cabbage worm, *Pieris rapae* L. after two days from treatment[53].

Mortalities are occurred in the 3rd day. The percent of mortalities are increased gradually and reached to the maximum in the 10th day from treatment.

The percent of mortalities ranged between 70 to 80, 60 to 75.7 and 40.0 to 55.2 % with *B. bassiana, M. anisopliae* and *V. lecanii*, respectively, in the 10th day after treatment. This means that *B. bassiana* isolation is more effective than *M. anisopliae* and *V. lecanii,* [53].

The % mortalities with all concentrations of *B. bassiana* isolation were 70, 72.4 and80 %, respectively.

Also, mortalities are occurred in the 3rd day. The percent of mortalities are increased gradually and reached to the maximum in the 10th day from treatment.

Data also showed a positive correlation between concentrations of fungi and the percentage of *P. rapae* L. Pupae mortality. The percent of mortalities ranged between 65 to 76, 58 to 70.1 and 42.0 to 60 % with *B. bassiana, M. anisopliae* and *V. lecanii*, respectively, in the 10th day after treatment. This means that *B. bassiana* isolation is more effective than *M. anisopliae* and *V. lecanii*. The percent of mortalities with all concentrations of *B. bassiana* isolation were 65, 70 and76 %, respectively [53].

Whitefly, *Bemisia tabaci* (Genn.)

There are no effect for *V. lecanii*, *M. anisopliae* and *B. bassiana* to *B. tabaci* after 3 days from treatment.

The percent of mortalities were increased gradually and reached to the maximum in the 7th day from treatment. With the all concentrations, the percent of mortalities were increased with increase of concentrations, [54].

The % mortalities ranged between 68.3 to 100, 56.2 to 90.5 and 62.7 to 97.3 % with *V. lecanii*, *M. anisopliae* and *B. bassiana*, respectively, in the 7th day after treatment. This mean that *V. lecanii* isolation is more effective than *M. anisopliae* and *B. bassiana*. The percent of mortalities with all concentrations of *V. lecanii* isolation were 68.3, 90.4and 100%, respectively, [54].

The number of *B. tabaci* per leave was decreased compared with control after the 2nd application. That the percent of reduction by *V. lecanii M. anisopliae* and *B. bassiana* after the third application were 57.3, 90.4 & 100, 50.2, 70.0 & 75.5 and 55.5, 79.5 & 90 % with C_1, C_2 and C_3, respectively[54].

There were no significant differences between the concentrations and controls after the first application all plots. After the 3rd application there were highly significant differences among all concentrations. The L.S.D was two, three after the third application. The statistical analysis confirmed that the third concentration (1 x 10^9) was the highly toxic compared the first and the second concentrations in all treatment, [54].

The percent of reduction by *V. lecanii M. anisopliae* and *B. bassiana* after the third application were 55.2, 85.2 & 100, 50.0, 71.0 & 76.2 and 56.2, 78.3 & 92 % with C_1, C_2 and C_3, respectively, [55].

The % of reduction by *V. lecanii M. anisopliae* and *B. bassiana* after the third application were 55.2, 85.2 & 100, 50.0, 71.0 & 76.2 and 56.2, 78.3 & 92 % with C_1, C_2 and C_3, respectively, [54].

Rhyzopertha dominica, Sitophilus oryzae and *Oryzaephilus surinmensis.*

R. dominica LC_{50} and LC_{90} values confirmed that *R. dominica* was more susceptible to *M. anisopliae* than *B. bassiana*, where LC_{50} and LC_{90} were 1.2 x 10^5, 2.3 x 10^5 conidia /g and 2.7 x 10^5, 0.01 x 10^6 conidia /g, respectively. The LT_{50}s were calculated as 4.6 and 6.6 days for the two respective fungi, indicating the superiority of *B. bassiana* over *M. anisopliae*, [53].

Sitophilus oryzae

The % cumulative mortality of *R. dominica* maximum mortality percentages adult were recorded 11 days after treatment with *B. bassiana* and *M. anisopliae*, where the mortality percentages were 25.2, 38.1, 45.2 and 50.3% for *B. bassiana* at the tested concentrations of 0.12 x 10^6conidia /g, 0.22 x 10^6conidia /g, 0.32 x 10^6conidia / g, and 0.42 x 10^6 conidia /g, respectively, and 63.0, 72.3, 75.3, and 79.3% for *M. anisopliae* at the tested concentrations of 0.12 x 10^6conidia /g, 0.22 x 10^6conidia /g, 0.32 x 10^6conidia / g, and 0.42 x 10^6 conidia / g, respectively.

The cumulative mortality percentages started with low levels on the 3rd day after treatment. Then the Percent mortality of *S. oryzae* was increased gradually, till the maximum was recorded on the 11th days of observation, where the mortality percentages were recorded 23, 32, 40 and 54% for *B. bassiana* respectively. The *S. oryzae* was more susceptible to infection with *B. bassiana* than *M. anisopliae.* LC_{50} and LC_{90} values proved that *S. oryzae* was more susceptible to *M. anisopliae* than *B. bassiana*, where the LC_{50} and LC_{90} of, *B. bassiana* and *M. anisopliae* were 4.3 x 10^5 and 0.08 x 10^6 conidia /g, and 1.6 x 10^5 and 7.1 x 10^5conidia /g, respectively.

LT_{50}s were calculated as 8.8 and 6.3 days for the two respective fungi, indicating that the conidial spores of *B. bassiana* were less efficient on *S. oryzae* than those of *M. anisopliae,* [55].

Oryzaephilus surinmensis

The same effective was found when the entomopathogenic fungi were used against *O. surinmensis*. The concentrations 0.42×10^{6} conidia / g of fungi, induced increase the mortalities rates.

The highest cumulative mortalities were recorded after 11 days of treatment when mortality percentages were recorded 25, 35, 43 and 51% for *B. bassiana* at the tested concentrations of (0.12×10^{6}conidia /g, 0.22×10^{6}conidia /g, 0.32×10^{6}conidia / g, and 0.42×10^{6} conidia / g, respectively).

The mortality percentage of *O. surinmensis* treated with *B. bassiana* and *M. anisopliae* differed significantly among different concentrations. Also mortality percentage differed significantly between *O. surinmensis* treated with *B. bassiana* and with *M. anisopliae,* [55].

Application in the field of isolated fungi against Many Insects

S. ocellatella* and *C. Vittata

The application of *B. bassiana* and *M. anisopliae* blast spores in sugar – beet fields was applied in Kafr El- Sheikh (Abu Kalab), Egypt. Sugar – beet plants were sprayed with the fungal suspensions to control *C. vittata*. Blastspores were grown on agitated bio malt liquid medium.

Blastspores of *B. bassiana* and *M. anisopliae* were applied to sugar- beet plants by using (10) lit. Knapsack sprayer with a concentration of (10^8 spores / ml.). Another untreated plot was used as control. In the treated and control plots, eggs, larvae, pupae and adults were counted on (25) plants .The spraying was repeated for three times in week, three weeks and five weeks, [60].

The field treatment of *C. vittata* with *M. anisopliae*

Using of *M. anisopliae* in the field application against the insect of *C. vittata*. Number of eggs, larvae, pupa and adult was counted and recorded in treatment and control before the beginning of the treatment and every two weeks after the application.

Before spraying the average number of eggs in treatment was 0.6 /plant and 0.26 / plant in control, the average number of larvae in treatment was 1.12/ plant and 1.2 / plant in control, the average number of pupa in treatment was 0 / plant and 0 / plant in control, the average number of adult in treatment was 0.6/ plant and 0.6 /plant.

After two weeks from the application the average number of eggs in treatment were 2.64/ plant and in control was 5.76 egg / plant, the average number of larvae in treatment was 3.32 /plant and in control was 5.24 / plant, average number of pupae in treatment was 2.0 / plant and in control was 2.08 /

plant, the average number of adult in treatment was 3.28 / plant and in control was 4.28 / plant.

After 4 weeks after treatment the average number of eggs in treatment was 1.84 / plant and in control was 3.44 / plant, the average number of larvae in treatment was 3.96 / plant and in control was 7.6 / plant, the average number of pupae in treatment was 1.12/plant and in control was 1.64/plant, the average number of adult in treatment was 2.36 / plant and in control was 4.84 /plant

After six weeks after treatment the average number of eggs was 4.84 / plant in treatment and 4.92 / plant in control, the average number of larvae was 5.52/ plant in treatment and 10.4 / plant in control, the average number of pupae was 0.92 / plant in treatment and 1.5/ plant in control, the average number of adult was 1.68 / plant in treatment and 2.92 /plant in control, [60].

Table (9) Average number of *C. vittata* different stages treatment with the entomopathogenic fungus *M. anisopliae,*[60]

Experimental Time	Average number of *C. vittata* / 25 plants							
	Eggs		Larvae		Pupae		Adults	
	T	C	T	C	T	C	T	C
Before First spray	0.6	0.68	1.12	1.2	0	0	0.6	0.6
Before Second spray	2.64	5.76	3.32	5.24	2.0	2.08	3.28	4.28
Before Third spray	1.84	3.44	3.96	7.6	1.12	1.64	2.36	4.84
After the Third spray	4.84	4.92	5.52	10.4	0.92	1.5	1.68	2.92

The field treatment of *C. vittata* with *B. bassiana:*

Using of *B. bassiana* in the field application against the insect of *C. vittata*. Number of eggs, larvae, pupa and adult was counted and recorded in treatment and control before the beginning of the treatment and every two weeks after the application.

Before spraying the average number of eggs in treatment was 6.0 /plant and 5.8 / plant in control, the average number of larvae in treatment was 4.8/ plant and 5.2 / plant in control, the average number of pupa in treatment was 1.9 / plant and 2.0 / plant in control, the average number of adult in treatment was 4.6/ plant and 4.2/plant in control, [60].

After two weeks from the application the average number of eggs in treatment were 1.88/ plant and in control was 3.44 egg / plant, the average number of larvae in treatment was 3.2 /plant and in control was 7.6 / plant, average number of pupae in treatment was 0.88 / plant and in control was 1.6 / plant, the average number of adult in treatment was 2.6 / plant and in control was 4.8/ plant, [60].

After four weeks after treatment the average number of eggs in treatment was 3.24 / plant and in control was 4.92/ plant, the average number of larvae in treatment was 4.2 / plant and in control was 10.4 / plant, the average number of pupae in treatment was 1.04/plant and in control was 1.5/plant, the average number of adult in treatment was 1.6 / plant and in control was 2.9 /plant, [60].

After 6 weeks after treatment the average number of eggs was 1.8 / plant in treatment and 3.64 / plant in control, the average number of larvae was 2.84/ plant in treatment and 7.7/ plant in control, the average number of pupae was 3.2 / plant in treatment and 7.2/ plant in control, the average number of adult was 2.7/ plant in treatment and 7.9/plant in control, [60].

Table (10): Average number of *C. vittata* different stages treatment with entomopathogenic fungus *B. bassiana*, [60]

Experimental Time	Average number of *C. vittata* / 25 plants							
	Eggs		Larvae		Pupae		Adults	
	T	C.	T	C	T	C	T	C
Before First Spray	6.0	5.8	4.8	5.2	1.9	2.0	4.6	4.2
Before 2nd spray	1.88	3.44	3.2	7.6	0.88	1.6	2.6	4.8
Before 3rd Spray	3.24	4.92	4.2	10.4	1.04	1.5	1.6	2.9
After 4th Spray	1.8	3.64	2.84	7.7	3.2	7.2	2.7	7.9

Table (11): Average number of *C. vittata* different stages treated with the entomopathogenic fungi *M. anisopliae,* [60]

Experimental Time	Average number of *C. vittata* / plants							
	Eggs		Larvae		Pupae		Adults	
	T	C	T	C	T	C	T	C
First week	0.6	0.68	1.12	1.2	0	0	0.6	0.6
Second week	2.64	5.76	3.32	5.24	2.0	2.08	3.28	4.28
Third week	1.84	3.44	3.96	7.6	1.12	1.64	2.36	4.84
Fourth week	4.84	4.92	5.52	10.4	0.92	1.5	1.68	2.92

Table (12): Average number of *C. vittata* different stages treated withentomopathogenic fungi *B. bassiana,* [60]

Experimental Time	Average number of *C. vittata* / plants							
	Eggs		Larvae		Pupae		Adults	
	T	C.	T	C	T	C	T	C
First week	6	5.8	4.8	5.2	1.9	2.0	4.6	4.2
Second week	1.88	3.44	3.2	7.6	0.88	1.6	2.6	4.8
Third week	3.24	4.92	4.2	10.4	1.04	1.5	1.6	2.9
Fourth week	1.8	3.64	2.84	7.7	3.2	7.2	2.7	7.9

Table (13) Average number of *C. vittata* different stages treatment with the entomopathogenic fungi *M. anisopliae*, [60]

Experimental Time	Average number of *C. vittata* / 25 plants							
	Eggs		Larvae		Pupae		Adults	
	T	C	T	C	T	C	T	C
First week	0.6	0.68	1.12	1.2	0	0	0.6	0.6
Second week	2.64	5.76	3.32	5.24	2.0	2.08	3.28	4.28
Third week	1.84	3.44	3.96	7.6	1.12	1.64	2.36	4.84
Fourth week	4.84	4.92	5.52	10.4	0.92	1.5	1.68	2.92

Table (14): Average number of *C. vittata* different stages treatment withentomopathogenic fungi *B. bassiana*, [60]

Experimental Time	Average number of *C. vittata* / 25 plants							
	Eggs		Larvae		Pupae		Adults	
	T	C.	T	C	T	C	T	C
First week	6	5.8	4.8	5.2	1.9	2.0	4.6	4.2
Second week	1.88	3.44	3.2	7.6	0.88	1.6	2.6	4.8
Third week	3.24	4.92	4.2	10.4	1.04	1.5	1.6	2.9
Fourth week	1.8	3.64	2.84	7.7	3.2	7.2	2.7	7.9

Table (15): Efficacy of the Fungi isolations against C. *vittata* larvae in sugar beet field, [60].

Fungi isolation	Conc. (ml)	Mean No. of alive larvae / plant & % Reduction after 4th applications during four weeks									Avg. % Red.
		1st spray			2nd spray		3rd spray		4th spray		
		Population before 1st spray	Population after one week	% Red.	Population after two week From spray	% Red.	Population after three weeks	% Red.	Population after four weeks	% Red.	
B. bassiana (B_6As)	0.5	4.0a	3.0b	36.0	2.2b	64.7	2.0b	52.7	1.6b	84.0	59.35
	1.0	3.8a	2.2bcd	51.0	1.6b	73.0	1.2bc	83.8	1.0b	88.2	74.0
	2.0	3.8a	1.4cde	68.5	1.0b	83.1	0.8c	89.8	0.6b	92.9	83.6
M. anisopliae (M_4As)	0.5	4.0a	2.4bc	48.6	2.0b	67.9	1.4bc	82.0	1.2b	86.5	71.3
	1.0	4.0a	1.2de	74.3	1.2b	80.8	0.8c	89.8	0.4b	95.5	85.1
	2.0	3.8a	0.8e	82.0	0.8b	86.5	0.4c	94.6	0.2b	97.3	90.1
control	0.0	3.6a	4.2a	-----	5.6a	-----	7.0a	-----	8.0a	-----	------
F		0.02 un	8.9*		10.87*		38.9*		21.5*		------
$LSD_{0.05}$		2.65	1.13		1.44		1.65		1.7		------

Treatment means followed by the same letter are not significantly different from each other ($P < 0.05$).

Table (16): Efficacy of the Fungi isolations against C. *vittata* larvae in sugar beet field,[60].

Fungi isolation	Conc. (ml)	Mean No. of alive larvae / plant & % Reduction after 4th applications during four weeks									Avg. % Red.
		1st spray			2nd spray		3rd spray		4th spray		
		Population before 1st spray	Population after one week	% Red.	Population after two week From spray	% Red.	Population after 3 weeks	% Red.	Population after 4weeks	% Red.	
B. bassiana (B6As)	0.5	3.0a	2.6ab	41.0	2.0b	56.7	1.6b	69.9	1.0b	85.1	63.2
	1.0	3.0a	2.0ab	55.0	1.6bc	65.4	0.6c	88.7	0.6b	91.1	75.8
	2.0	3.0a	1.6ab	64.0	1.0bc	78.4	0.4c	92.5	0.4b	94.1	82.25
M. anisopliae (M4As)	0.5	3.0a	2.0ab	54.4	1.0bc	78.4	0.6c	88.7	0.8b	88.1	77.4
	1.0	3.0a	1.6ab	63.6	0.8c	82.7	0.6c	88.7	0.4b	94.1	82.3
	2.0	2.6a	1.2ab	68.5	0.6c	85.7	0.4bc	91.4	0.2b	96.6	85.6
control	0.0	2.6a	3.8a	-----	4.0a	-----	4.6a	----	5.8a		----
F		0.08un	2.96un		10.05*		19.7*		18.42*		
$LSD_{0.05}$		1.93	1.45		1.07		0.17		1.09		

Treatment means followed by the same letter are not significantly different from each other ($P < 0.05$)

The Potato tuber moth, *Phathorimaea operculella* (ZELLER)

Application of the fungi occurred was accomplished at the rate of 1X 10^8 spores/ml. Applications were made by a sprayed at the sunset. Four applications were made at 4- weeks intervals during crop growing season. Control plots were left without any treatments. Examinations of 40 plants /plot /treatment were carried out just before the first application and seven days after last application to calculate the average reduction percentages in the target insect infestation percentages which were calculated in each treatment. The agricultural practices followed the recommendations of the Ministry of Agricultural. Twenty tubers were taken from the first 5 rows in each treatment and in the control, [53].

Under field conditions the yields weight of potatoes were significantly increased to 25.45± 55.66 and 29.67±61.11 ton/ feddan in plots treated with *B. brongniartii* and *N. rileyi* respectively as compared to17.88±55.43 Tons/ feddan in the control. The weight of potatoes were significantly increased to 25.97±67.91 and 29.94±54.98 Ton/ feddan as compared to 12.27±45.09 in the control, [53].

Sitophilus granarius

The effect of the grain weevil *S. granarius* were significantly decreased after oil extract treatments , the percentage of larval mortality , 88.14, 39.13 and 2.21 after *N. Sativa* treatments at the concentrations of 5, 0.5 and 0.05%. The corresponding concentrations of *T. distichum* gave the larval mortality 57.18, 20.26 and 4.41 % of the *S. granarius,* respectively. When *B. carterii* treated with the last concentrations the larval mortality were significant decrease to 29.61, 15.81 and 0.613% as compared to zero in the control, [55].

The oviposition deterrent effect in the means number of eggs were significantly decreased to 13.0±7.8‚ 37±7.9 and 45±3.8 eggs/ female as when treated with 3% of the oil *N. sativa , T. distichum and B. carterii*, respectively compared to 198±4.9 in the control,[55].

The half-life period of the target insect pest after oil extract treatments which show that the LC_{50} obtained 2.1 X 10^{5}‚ 3.2 X 10^{5}‚ and 3.7 X 10^{5} as compared to 8.8 X 10^{5} sores /ml. [55].

Accumulative mortality of *S. granarius* during the first week of rice seeds exposed to treated foam with *N. sativa, T. distichum,* and *B. carterii* on *S. granarius* oils. After seven days from treated rice seeds the accumulative mortality was 78.15, 50.17, and 30.1, respectively compared with 2.1 in untreated, [55].

Pieris rapae

There are significant difference between 1st (C_1) and 2nd (C_2) spores concentrations and control after the first application in all parts, the differences appear gradually after the second and third application. On the other hand the third concentration (C_3) in *B. bassiana* was the best concentration against *P. rapae* L. followed by the third concentration in *M. anisopliae* and the third concentration in *V. lecanii*. This means that the third concentration in *B. Bassiana* was the best concentrations against *P. rapae,* [53].

There are significant difference also between 1st (C_1) and 3rd (C_2) spores concentrations and control after the first application in all parts, the differences appear gradually after the second and third application. On the other hand the third concentration (C_3) in *B. bassiana* was the best concentration against *P. rapae* followed by the third concentration in *M. anisopliae* and the third concentration in *V. lecanii*. The Percent of reduction in all treatment was ranged between 48.5 to 75.8% in all concentrations, [53].

B. Bassiana, *M. anisopliae* and *V. lecanii* isolates are promising agents for Cabbage worm control in the field, [53].

Sugar beet fly, *Pegomyia mixta*

P. mixta started to appear during the second week of January. In mid-February, the first application was carried out using two entomopathogenic fungi *B. bassiana,* and *M. anisopliae,* [57].

The number of infested / 10 plants was 70 % before the first spray by *B. bassiana* and reached to 10 % after seven days from treatment and reached to zero% infested after 14 day from first treatment. Also, the number of infested leaves 11 before the first spray by *B. bassiana* and reached to 4 after seven days from treatment and reached to zero infested after 14 day from first treatment, [57]. The number of infested egg patches / plant 13 before the first spray by *B. bassiana* and reached to 5 after seven days from treatment and reached to zero infested after 14 day from first treatment.

The number of infested larvae / plant 22 before the first sprays by *B. bassiana* and reached to 4 after seven days from treatment and reached to zero infested after 14 day from first treatment.

The number of infested / 10 plants was 80 % before the first spray by *M. anisopliae* and reached to 20 % after seven days from treatment and reached to 10% infested after 14 day from first treatment. Also, the number of infested leaves 12 before the first spray by *B. bassiana* and reached to 6 after seven days from treatment and reached to 2% infested after 14 day from first treatment.

The number of infested egg patch / plant 12 before the first spray by *B. bassiana* and reached to 6 after seven days from treatment and reached to 3% infested after 14 day from first treatment.

The number of infested larvae / plant 21 before the first spray by *B. bassiana* and reached to 15 after seven days from treatment and reached to zero infested after 14 day from first treatment, [57].

The number of infested / 10 plants 60 % before the first spray by *B. bassiana* and reached to 20 % after seven days from treatment and reached to zero% infested after 14 day from first treatment. Also, the number of infested leaves 13 before the first spray by *B. bassiana* and reached to 3 after seven days from treatment and reached to zero infested after 14 day from first treatment.

The number of infested egg patches / plant 12 before the first spray by *B. bassiana* and reached to 4 after seven days from treatment and reached to zero infested after 14 day from first treatment, [57].

The number of infested larvae / plant 25 before the first spray by *B. bassiana* and reached to 12 after seven days from treatment and reached to zero infested after 14 day from first treatment.

The number of infested / 10 plants 70 % before the first spray by *Metarhizium anisopliae* and reached to 30 % after seven days from treatment and reached to Zero% infested after 14 day from first treatment. Also, the number of infested leaves 13 before the first spray by *Metarhizium anisopliae* and reached to 6 after seven days from treatment and reached to Zero% infested after 14 day from first treatment.

The number of infested egg patches / plant 12 before the first spray by *M. anisopliae* and reached to 6 after seven days from treatment and reached to Zero% infested after 14 day from first treatment.

The number of infested larvae / plant 23 before the first sprays by *M. anisopliae* and reached to 14 after seven days from treatment and reached to zero infested after 14 day from first treatment.

The entomopathogenic fungi *B. bassiana* and *M. anisopliae* can be used as a promising agent in pest control and integrated pest management programs instead of conventional pesticides to reduce the environmental pollution especially when the pests were under the economic threshold, [57].

The Oliver black scale insect, *Saissetia oleae* (Oliver)

Percent of sampled substrate infested with black scale Insect, *S. oleae* (Olivier) before and after treatment with *B. bassiana* reduction from 92.0% in the first week from treatment to 9.2 % after seven weeks from treatment, [58].

Also when we treated the trees with *M. anisopliae* reduction in population of the black scale Insect, *S. oleae* was 2.1% from 93.0% after seven weeks from treatment, [58].

Also when we treated the trees with *V. lecanii* reduction in population of the black scale Insect, *S. oleae* was 9.5% from 90.0% after seven weeks from treatment. *M. anisopliae* caused the highest reduction in black scale Insect, *S. oleae* than *B. bassiana* and *V. lecanii*, [58].

Percent of sampled substrate infested with black scale Insect, *S. oleae* (Olivier) before and after treatment with *B. bassiana* reduction from 84.0% in the first week from treatment to 8.3 % after seven weeks from treatment, , [58].

Also when we treated the trees with *M. anisopliae* reduction in population of the black scale Insect, *S. oleae* was 1.0 % from 86.0% after seven weeks from treatment.

Also when we treated the trees with *V. lecanii* reduction in population of the black scale Insect, *S. oleae* was 8.5% from 85.0% after seven weeks from treatment. *M. anisopliae* caused the highest reduction in black scale Insect, *Saissetia oleae* than *B. bassiana* and *V. lecanii*, [58].

Anti-soft scale insect *Pulvinaria tenuivalvata*

The mean numbers of adults found on leaves before treatments, ranged from 58.6 to 68.2scales / 50 leaves, indicating a relatively uniform distribution of insect infestation. Three days after spraying, the treatments suppressed the levels of infestation to different degrees as compared to untreated control. *Beauveria bassiana* (10^8 spores/ml), Petroleum ether fraction of *Cressa cretica* with a rate of (50ml/L.) and significantly lowered the percentage of infestation to 39.9% and 44.6 %, respectively, although they didn't reach KZ oil (15ml/L.) activity (54.0%). Two weeks after treatments, *B. bassiana* (10^8 spores/ml), Petroleum ether fraction (50ml/L.) and KZ oil (15ml/L.) had almost similar activity, 65.8, 69.1 and 79.4% reduction in the infestation, respectively.

Statistical analysis showed significant differences between tested compounds and control on the soft scale insect population. The low susceptibility of the adult stage to the tested compounds may be attributed to the presence of protective scales which prevent the penetration of sprayings, [59].

Increased percentages over the control ranged between 1.23 to 1.27 when used *Beauveria bassiana*, *Cressa cretica* Petroleum ether fraction 5% and KZ oil 95%, respectively. The average calculated yield in the control was 21710 Kg / feddan, as compared to 26750, 27180 and 27670 Kg / feddan after using the *Beauveria bassiana*, *Cressa cretica* Petroleum ether fraction 5% and KZ oil 95%, respectively, [59].

The green stink bug, *Nezara viridula* L. in sugar-beet

Table 6 showed that the third concentration (3×10^5) in *Metarhizium anisopliae* was the best concentration against *Nezara viridula* followed by the third concentration in *B. bassiana* isolation. The same results were found by El-Zoghby (2003) and Abdel-Raheem (2009 and 2010).

After the second application the average number of *Nezara viridula* decreased compared with control. That the average numbers by *Metarhizium anisopliae* after the third applications were 2.0, 1.0 and 0.0 with 3×10^3, 3×10^4 and 3×10^5, respectively and the average numbers by *B. bassiana* after the third applications were 4.0, 2.0 and 1.0 with 3×10^3, 3×10^4 and 3×10^5, respectively, **[61]**

During 2010 season Table 7 showed that the third concentration (3×10^5) in *Metarhizium anisopliae* was the best concentration against *Nezara viridula* followed by the third concentration in *B. bassiana* isolation. The same results were found by El-Zoghby (2003) and Abdel-Raheem (2009 and 2010).

After the second application the average number of *Nezara viridula* decreased compared with control. That the average numbers by *Metarhizium anisopliae* after the third applications were 1.0, 1.0 and 0.0 with 3×10^3, 3×10^4 and $3x\ 10^5$, respectively and the average numbers by *B. bassiana* after the third applications were 7.0, 3.0 and 2.0 with 3×10^3, 3×10^4 and 3×10^5, respectively, **[61]**

Finally, these data clear that the entomopathogenic fungi *Metarhizium anisopliae* and *B. bassiana* can be used as a promising agent in pest control and integrated pest management programs instead of conventional pesticides to reduce the environmental pollution especially when the pests were under the economic threshold. The entomopathogenic fungi *Metarhizium anisopliae* was more promising than *B. bassiana* on *Nezara viridula*, **[61]**

Table (17): Average number of *Nezara viridula* on sugar-beet plants before spraying the two entomopathogenic fungi during Season (2009), [61]

Before Spray	Cont.	*M. anisopliae*			*B. bassiana*		
		3 x 10^3	3 x 10^4	3x 10^5	3 x 10^3	3 x 10^4	3x 10^5
1st	9	8	9	9	10	8	7
2nd	12	5	5	4	9	7	5
3rd	10	3	2	2	7	6	3
4th	5	2	1	0	4	2	1

Table (18): Average number of *Nezara viridula* on sugar-beet plants before spraying the two entomopathogenic fungi during season (2010), [61]

:

Before Spray	Cont.	*M. anisopliae*			*B. bassiana*		
		3 x 10^3	3 x 10^4	3x 10^5	3 x 10^3	3 x 10^4	3 x 10^5
1st	13	10	12	10	12	13	11
2nd	10	7	7	5	10	9	7
3rd	14	3	4	2	9	6	4
4th	7	1	1	0	7	3	2

Application of Entomopathogenic Fungi as Nano

The Red Palm Weevil, *Rhynchophorus ferrugineus* Olivier (Coleoptera: Curculionidae)

Data of the treated eggs of *R. ferrugineus* with *M. anisopliae*, *B. bassiana*, Bio Magic, Bio Power and Bio Catch as fungal spores and their Silver NPs particles was presented in table (19). Seven days post treatment, up- to (90%) mortality rate of *R. ferrugineus* was recorded in the treated eggs.[62]

The % mortality rates attained 80, 73, 65, 60, and 45 % by infection with fungal spores from *M. anisopliae*, *B. bassiana*, Bio Magic, Bio Power and Bio Catch, respectively.

Furthermore, the % mortality rates were 90, 84, 73, 70, and 58% by infection with bio synthesized Ag NPs from *M. anisopliae*, *B. bassiana*, Bio Magic, Bio Power and Bio Catch, respectively during the same time. *M. anisopliae* recorded the highest mortality (90%) in the eggs of *R. ferrugineus,* when treated with Nano-particles or with fungal spores after 6 days and was the lowest (58%) when treated with Bio Catch. According to Abdel-Raheem *et al.* [51] the total mortality of eggs and larvae were reduced than the control group when eggs were exposed to *M. anisopliae* spores.

(Abdel-Raheem, [62]) mentioned that the reason of different pathogenicity rates between one fungus and another may due to the fraction exhibiting antimicrobial activity of some polar compounds ranging between 1000 and 1500 Dalton in the extraction of fungi.

Table (19): Mortality percent of entomopathogenic fungi on the eggs of *Rhynchophorus ferrugineus,* using spore suspension and bio synthesized silver nanoparticles [62]

Entomopathogenic fungus	Treated with	
	Fungal spores (Mean ± S.E)	Bio synthesized Ag NPs (Mean ± S.E)
Metarhizium anisopliae	80.0 ± 1.20	90.0 ± 2.10
Beauveria bassiana	73.0 ± 1.00	84.0 ± 1.30
Bio Magic	65.0 ± 0.20	73.0 ± 0.10
Bio Power	60.0 ± 3.10	70.0 ± 0.20
Bio Catch	45.0 ± 0.20	58.0 ± 1.20
Control	6.0	7.0
S.E (m)	1.14	1.98

Data of larval mortality of *R. ferrugineus,* treated with *M. anisopliae*, *B. bassiana*, Bio Magic, Bio Power and Bio Catch as fungal Spores and Silver Nano-particles was presented in table (20). Seven days post treatment, 90% mortality of *R. ferrugineus* larvae was recorded as: 84, 75, 71, 65, and 55 % by infection with fungal spores from *M. anisopliae*, *B. bassiana*, Bio Magic, *Bio* Power and Bio Catch, respectively. By infection with bio synthesized Ag NPs from *M. anisopliae*, *B. bassiana*, Bio Magic, Bio Power and Bio Catch, the mortality rates recorded were 95, 87, 77, 73, and 60 %, respectively. *M. anisopliae* was the highest % mortality (95%) in the larvae of *R. ferrugineus,* when treated with the Nano-particles or with the fungal spores after 6 days, while the lowest (60%) was when treated with Bio Catch, *V. lecanii*. The results agree with Tefera and Pringle (2003) who reported that the bio efficacy of *M. anisopliae* in all stages of *R. ferrugineus* caused up to 48 to 95 % mortality of adult and larvae. [62]

Table (20): Mortality percent of entomopathogenic fungi on the larvae of *Rhynchophorus ferrugineus* using spore suspension and bio synthesized silver nanoparticles [62]

Entomopathogenic fungus	Treated with	
	Fungal spores (Mean ± S.E)	Bio synthesized Ag NPs (Mean ± S.E)
Metarhizium anisopliae	84.0 ± 2.10	95.0 ± 2.30
Beauveria bassiana	75.0 ± 2.12	87.0 ± 2.22
Bio Magic	71.0 ± 2.20	77.0 ± 1.10
Bio Power	65.0 ± 1.10	73.0 ± 1.20
Bio Catch	55.0 ± 0.10	60.0 ± 1.00
Control	6.2	7.3
S.E (m)	2.10	3.00

Data of adult mortality of *R. ferrugineus,* treated with *M. anisopliae*, *B. bassiana*, Bio Magic, Bio Power and Bio Catch as fungal Spores and Silver Nano-particles was presented in table (21). Seven days post treatment, the parentage of mortality of *R. ferrugineus* adults reached 77%. The % mortality rates recorded 65, 61, 56, 52, and 35 % by infection with fungal spores from *M. anisopliae*, *B. bassiana*, Bio Magic, *Bio* Power and Bio Catch, respectively. While, by infection with bio synthesized Ag NPs from *M. anisopliae*, *B. bassiana*, Bio Magic, Bio Power and Bio Catch, the % mortality recorded were 77, 70, 63, 55, and 48 %, respectively. *M. anisopliae* was the highest % mortality (77%) in the adults of *R. ferrugineus,* when treated with the Nano-particles or with fungal spores after 6 days, while the lowest was (48%), when they were treated by Bio Catch.

Table (21): Mortality percent of entomopathogenic fungi on the adults of *Rhynchophorus ferrugineus* using spore suspension and bio synthesized silver nanoparticles [62]

Entomopathogenic fungus	Treated with	
	Fungal spores (Mean ± S.E)	Bio synthesized Ag NPs (Mean ± S.E)
Metarhizium anisopliae	65.0 ± 2.00	77.0 ± 1.30
Beauveria bassiana	61.0 ± 1.20	70.0 ± 2.10
Bio Magic	56.0 ± 2.20	63.0 ± 0.10
Bio Power	52.0 ± 1.20	55.0 ± 3.20
Bio Catch	35.0 ± 0.00	48.0 ± 0.20
Control	7.0	8.0
S.E(m)	1.94	2.90

The Ag NPs synthesized through application of spore suspension and bio synthesized silver Nano-particles showed an efficacy against *R. ferrugineus* different stages. *M. anisopliae* had the highest potential and was more effective than all the others.

Virulence of Nano – Particle preparation of Entomopathogenic fungi and Entomopathogenic Bacteria against red palm weevil *Rhynchophorus ferrugineus* (Olivier) (Coleoptera: Curculionidae).

Rhynchophorus ferrugineus is one of the most severe pests on palm species, including date palms In Asia. The purpose of the current study is to evaluate the virulence of bio efficacy of nano - particle of entomopathogenic fungi and entomopathogenic bacteria on red palm weevil, *Rhynchophorus ferrugineus.* Prepare concentrations of Fungal Spores and Silver Nano Particles were prepared from *Metarhizium anisopliae*, *Beauveria bassiana, Verticillium lecanii* and *B. thuringiensis*, *B. subtilis.* [63]

M. anisopliae was the highest % mortality (90%) ,(95%) and (77 %) against eggs, larvae and adults of *R. ferrugineus* , respectively when treated with bio synthesized or with fungal spores after seven days from treatment and the lowest % mortality was recorded when treated with *V. lecanii.* [63]

Three concentrations, (10^3, 10^4, and 10^5 CFU/mL), formulated as bacterial suspensions and bio-synthesized silver nanoparticles, (Ag NPs) were evaluated for their ability to inhibit egg hatch. The 10^5 CFU/mL concentration, both as a bacterial suspension and formulated as Ag NPs, was tested for efficacy against 10 day-old larvae and adults. [63]

Egg hatch was significantly inhibited by all concentrations and formulations of *B. subtilis* and *B. thuringiensis*, and exhibited lethal concentrations for 50% mortality (LC_{50}) of 3.45 X 10^3 and 6.73 X 10^4 CFU/mL, respectively. The percent mortality of larvae six days after treatment was 85% and 77% for *B. thuringiensis* and *B. subtilis* Ag NPs, respectively. The percent mortality of adults six days after treatment was 75% and 67% for *B. thuringiensis* and *B. subtilis* Ag NPs, respectively. [63]

The efficacy of nanoparticle preparations of *Metarhizium anisopliae*, *Beauveria bassiana,* and *Verticillium lecanii* against eggs, larvae, and adults of the red palm weevil. [63]

Table (22 shows that the results of the eggs of *R. ferrugineus* which were treated with *M. anisopliae*, *B. bassiana*, Bio Magic, Bio Power and Bio Catch as fungal Spores and Silver nano particles. After six days of treatment, the petri dishes plates were observed and the result was 90% mortality of *R. ferrugineus* eggs. Additionally, white, green and metallic muscardine were found on the dead eggs. The % mortality after six days, from treated subjects, recorded 80, 73, 65, 60, and 45 % by infection with fungal spores from *M. anisopliae*, *B. bassiana*, Bio Magic, Bio Power and Bio Catch, respectively. [63]

Furthermore, the % mortality after six days from treated subjects recorded 90, 84, 73, 70, and 58 % by infection with bio synthesized Ag NPs from *M. anisopliae*, *B. bassiana*, Bio Magic, Bio Power and Bio Catch, respectively. *M. anisopliae* was the highest % mortality (90%) on the eggs of *R. ferrugineus* when treated with bio synthesized or with fungal spores after six days from treatment and the lowest % mortality was (58%) when treated with Bio Catch. [63]

Table (22): Bio efficacy of entomopathogenic fungi on the eggs of *R. ferrugineus* using spore suspension and bio synthesized silver nanoparticles.[63]

Entomopathogenic fungi	Treated with	
	Fungal spores (Mean ± S.E)	Bio synthesized Ag NPs (Mean ± S.E)
M. anisopliae	80.0 ± 1.20	90.0 ± 2.10
B. bassiana	73.0 ± 1.00	84.0 ± 1.30
Bio Magic	65.0 ± 0.20	73.0 ± 0.10
Bio Power	60.0 ± 3.10	70.0 ± 0.20
Bio Catch	45.0 ± 0.20	58.0 ± 1.20
Control	6.0	7.0
S.E(m)	1.14	1.98

Table (23) shows that the 4th larvae of *R. ferrugineus* were treated with *M. anisopliae*, *B. bassiana*, Bio Magic, Bio Power and Bio Catch as fungal Spores and Silver nano particles. After six days of treatment, the petri dishes plates showed up to 90% mortality of *R. ferrugineus* larvae. [63]

Also, white, green and metallic muscardine were found on the dead larvae. The % mortality after six days recorded 84, 75, 71, 65, and 55 % by infection with fungal spores from *M. anisopliae*, *B. bassiana*, Bio Magic, *Bio* Power and Bio Catch, respectively. Furthermore, the % mortality after six days recorded 95, 87, 77, 73, and 60 % by infection with bio synthesized Ag NPs from *M. anisopliae*, *B. bassiana*, Bio Magic, Bio Power and Bio Catch, respectively. *M. anisopliae* was the highest % mortality (95%) against the larvae of *R. ferrugineus* when treated with bio synthesized or with fungal spores after six days and the lowest % mortality was (60%) when treated with Bio Catch, *V. lecanii*. [63]

Table (23): Bio efficacy of entomopathogenic fungi on the 4th larvae of *R. ferrugineus* using spore suspension and bio synthesized silver nanoparticles.

Entomopathogenic fungi	Treated with	
	Fungal spores (Mean ± S.E)	Bio synthesized Ag NPs (Mean ± S.E)
M. anisopliae	84.0 ± 2.10	95.0 ± 2.30
B. bassiana	75.0 ± 2.12	87.0 ± 2.22
Bio Magic	71.0 ± 2.20	77.0 ± 1.10
Bio Power	65.0 ± 1.10	73.0 ± 1.20
Bio Catch	55.0 ± 0.10	60.0 ± 1.00
Control	6.2	7.3
S.E(m)	2.10	3.00

Table (24) shows that the adults of *R. ferrugineus* were treated with *M. anisopliae*, *B. bassiana*, Bio Magic, Bio Power and Bio Catch as fungal Spores and Silver nano particles.[63]

After six days of treatment, the petri dishes plates were observed and the parentage of mortality of *R. ferrugineus* adults was 77%. Also, white, green and metallic muscardine were found on the dead adults. The % mortality after six days recorded 65, 61, 56, 52, and 35 % by infection with fungal spores from *M. anisopliae*, *B. bassiana*, Bio Magic, Bio Power and Bio Catch, respectively. Moreover, the % mortality after six days of treatment recorded 77, 70, 63, 55, and 48 % by infection with bio synthesized Ag NPs from *M. anisopliae*, *B. bassiana*, Bio Magic, Bio Power and Bio Catch, respectively. *M. anisopliae* was the highest % mortality (77%) against the adults of *R. ferrugineus* when treated with bio synthesized or with fungal spores after six days and the lowest % mortality was (48%) when it treated with Bio Catch[63]

Table (24): Bio efficacy of entomopathogenic fungi on the adults of *R. ferrugineus* using spore suspension and bio synthesized silver nanoparticles. [63]

Entomopathogenic fungi	Treated with	
	Fungal spores (Mean ± S.E)	Bio synthesized Ag NPs (Mean ± S.E)
M. anisopliae	65.0 ± 2.00	77.0 ± 1.30
B. bassiana	61.0 ± 1.20	70.0 ± 2.10
Bio Magic	56.0 ± 2.20	63.0 ± 0.10
Bio Power	52.0 ± 1.20	55.0 ± 3.20
Bio Catch	35.0 ± 0.00	48.0 ± 0.20
Control	7.0	8.0
S.E(m)	1.94	2.90

Silver Nano-particles from Entomopathogenic Fungi on The tortoise beetle, *Cassida vittata* (Coleoptera: Chrysomelidae)

Data of the treated eggs of *C. vittata* with *M. anisopliae*, *B. bassiana*, Bio Magic, Bio Power and Bio Catch as fungal spores and their Silver NPs particles was presented in table (25).[64]

Seven days post treatment, up- to (95%) mortality rate of *C. vittata* was recorded in the treated eggs. The % mortality rates attained 84, 75, 68, 65, and 47 % by infection with fungal spores from *M. anisopliae*, *B. bassiana*, Bio Magic, Bio Power and Bio Catch, respectively. Furthermore, the % mortality rates were 95, 87, 75, 72, and 61% by infection with bio synthesized Ag NPs from *M. anisopliae*, *B. bassiana*, Bio Magic, Bio Power and Bio Catch, respectively during the same time. *M. anisopliae* recorded the highest mortality (95%) in the eggs of *C. vittata,* when treated with Nano-particles or with fungal spores after 6 days and was the lowest (61%) when treated with Bio Catch. According to [12, 51, 61-64] the total mortality of eggs and larvae were reduced than the control group when eggs were exposed to *M. anisopliae* spores.[64]

Table (25): % Mortality of EPF on the eggs of *C. vittata,* using spore suspension and bio synthesized silver nanoparticles [64]

EPF	Treated with	
	Fungal spores (Mean ± S.E)	Bio synthesized Ag NPs (Mean ± S.E)
M. anisopliae	84.0 ± 1.22	95.0 ± 2.11
B. bassiana	75.0 ± 1.10	87.0 ± 1.20
Bio Magic	68.0 ± 0.10	75.0 ± 0.11
Bio Power	65.0 ± 2.13	72.0 ± 1.11
Bio Catch	47.0 ± 1.12	61.0 ± 1.25
Control	5.0	5.0
S.E (m)	1.13	1.16

Data of larval mortality of *C. vittata,* treated with *M. anisopliae*, *B. bassiana*, Bio Magic, Bio Power and Bio Catch as fungal Spores and Silver Nano-particles was presented in table (26).

Seven days post treatment, 97% mortality of *C. vittata* larvae was recorded as: 85, 78, 72, 67, and 58 % by infection with fungal spores from *M. anisopliae*, *B. bassiana*, Bio Magic, *Bio* Power and Bio Catch, respectively.

By infection with bio synthesized Ag NPs from *M. anisopliae*, *B. bassiana*, Bio Magic, Bio Power and Bio Catch, the mortality rates recorded were 87, 87, 80, 76, and 64 %, respectively. *M. anisopliae* was the highest % mortality (97%) in the larvae of *C. vittata,* when treated with the Nano-particles or with the fungal spores after 6 days, while the lowest (64%) was when treated with Bio Catch, *V. lecanii*. [64]

Table (26): % Mortality of EPF on the larvae of *C. vittata* using spore suspension and bio synthesized silver nanoparticles [64]

EPF	Treated with	
	Fungal spores (Mean ± S.E)	Bio synthesized Ag NPs (Mean ± S.E)
M. anisopliae	85.0 ± 2.17	97.0 ± 1.11
B. bassian	78.0 ± 2.42	87.0 ± 2.22
Bio Magic	72.0 ± 1.13	80.0 ± 1.11
Bio Power	67.0 ± 1.22	76.0 ± 1.11
Bio Catch	58.0 ± 1.22	64.0 ± 1.10
Control	6.0	6.0
S.E (m)	1.63	1.33

Data of adult mortality of *C. vittata,* treated with *M. anisopliae*, *B. bassiana*, Bio Magic, Bio Power and Bio Catch as fungal Spores and Silver Nano-particles was presented in table (27). Seven days post treatment, the parentage of mortality of *C. vittata,* adults reached 80%.

The % mortality rates recorded 69, 65, 62, 58, and 55 % by infection with fungal spores from *M. anisopliae*, *B. bassiana*, Bio Magic, *Bio* Power and Bio Catch, respectively. While, by infection with bio synthesized Ag NPs from *M. anisopliae, B. bassiana*, Bio Magic, Bio Power and Bio Catch, the % mortality recorded were 80, 77, 69, 65, and 60 %, respectively.

M. anisopliae was the highest % mortality (80%) in the adults of *C. vittata,* when treated with the Nano-particles or with fungal spores after 6 days, while the lowest was (60%), when they were treated by Bio Catch.[64]

Table (27): % Mortality of EPF on the adults of *C. vittata* using spore suspension and bio synthesized silver nanoparticles [64]

EPF	Treated with	
	Fungal spores (Mean ± S.E)	Bio synthesized Ag NPs (Mean ± S.E)
M. anisopliae	69.0 ± 2.11	80.0 ± 1.22
B. bassiana	65.0 ± 1.20	77.0 ± 1.00
Bio Magic	62.0 ± 2.11	69.0 ± 1.10
Bio Power	58.0 ± 1.12	65.0 ± 3.20
Bio Catch	55.0 ± 1.11	60.0 ± 1.20
Control	6.0	7.0
S.E(m)	1.53	1.54

SILVER NANOPARTICICLES FROM ENTOMOPATHOGENIC FUNGI AGAINST THE SPINY BOLLWORM, *EARIAS INSULANA* IN MAIZE CROP

Data of the treated nymphs of *E. insulana* with *B. bassiana*, *M. anisopliae* and Bio Catch (*V. lecanii*) as fungal spores and their Silver NPs particles were presented in table (28). [65]

It appears that both treatments, fungal spores and bio-synthesized Ag NPs, have a positive impact on the EPF (Entomopathogenic Fungi) being studied, as compared to the control group.Six days post-treatment, up-to (70%) mortality rate of *E. insulana* was recorded in the treated larvae. The percent mortality rates attained 60, 50, and 47% by infection with fungal spores from *B. bassiana*, *M. anisopliae,* and *V. lecanii*, respectively. [65]

Furthermore, the percentage mortality rates were 70, 60, and 58% by infection with biosynthesized Ag NPs from *B. bassiana*, *M. anisopliae* and *V. lecanii* respectively, at the same time. *B. bassiana* recorded the highest mortality (70%) in the larvae of *E. insulana*, when treated with nanoparticles or with fungal spores after 6 days will the lowest was (47%) when treated with *V. lecanii* fungal spores. The control group, which did not undergo any treatment, maintained a constant value of 5.0. This suggests that there was no significant impact on the EPF's activity in the absence of treatment. [65]

The standard error values (S.E) provided in the last row of the table indicate the variability or margin of error associated with the mean values. The S.E for fungal spores treatment is 1.3, while for bio-synthesized Ag NPs treatment, it is 1.40. These values represent the range within which the true mean values are expected to fall. [65]

Table (28): Percent mortality of EPF on the larvae of *E. insulana,* using spore suspension and biosynthesized silver nanoparticles.[65]

EPF	Treated with	
	Fungal spores (Mean ± S.E)	Bio synthesized Ag NPs (Mean ± S.E)
B. bassiana	60.0 ± 2.3	70.0 ± 2.2
M. anisopliae	50.0 ± 1.20	60.0 ± 1.2
V. lecanii	47.0 ± 1.1	58.0 ± 1.1
Control	5.0	5.0
S.E (m)	1.3	1.40

The pupae percent of mortality of *E. insulana,* treated with *B. bassiana, M. anisopliae* and *V. lecanii* as fungal Spores and Silver nanoparticles was presented in table (2). Sex days post-treatment, mortality percent of *E. insulana* pupae was recorded: 57, 45, and 40 % by infection with fungal spores from *B. bassiana, M. anisopliae* and *V. lecanii,* respectively. [65]

Will, at the infection with biosynthesized Ag NPs from *B. bassiana, M. anisopliae* and *V. lecanii* the mortality rates recorded 65, 60, and 50%, respectively. *B. bassiana* was the highest % mortality (65%) in the pupae of *E. insulana,* when treated with the nanoparticles or with the fungal spores (57%) after 6 days, while the lowest (50%) was when treated with *V. lecanii* biosynthesized. The control group, as before, shows a constant value of 5.0, indicating no significant change or treatment effect. The standard error values (S.E) provided in the last row of the table indicate the variability or margin of error associated with the mean values. The S.E for the fungal spores treatment is 1.31, while for the bio-synthesized Ag NPs treatment, it is 1.51. These values represent the range within which the true mean values are expected to fall. [65]

Table (29): Percent mortality of EPF on the pupae of *E. insulana* using spore suspension and biosynthesized silver nanoparticles[65]

EPF	Treated with	
	Fungal spores (Mean ± S.E)	Bio synthesized Ag NPs (Mean ± S.E)
B. bassiana	57.0 ± 1.2	65.0 ± 1.11
M. anisopliae	45.0 ± 1.2	60.0 ± 1.11
V. lecanii	40.0 ± 1.2	50.0 ± 1.10
Control	5.0	5.0
S.E (m)	1.31	1.51

CONCLUSION

Entomopathogenic fungi are a specialized group of microbial organisms with traits and modes of action that render them effective as biopesticides. Hypocreales species are ubiquitous parts of the soil microbiota. Soil is important for protection against environmental extremes, for water availability, and, also as a source of insect hosts.

While the chemical pesticides are becoming obsolete, the urge for research and development of biopesticides is growing. While only a few species of *Beauveria*, *Metarhizium*, *Lecanicillium* and *Isaria* are commercially available, the recent surge in studies exploring the endophytical attributes of fungi paves the way to further research into their complex ecology.

Acknowledging their ecology better, biopesticides can potentially develop to more efficient and more diversified applications. Better grasp of their compatibility with other microbial agents, natural arthropod enemies, or plant and/or chemical compounds could generate a richer and stronger pool of available tools for pest control. Also, research should be focused on discovering the most appropriate strains or isolates per plant host. Differential rates of endophytism and degrees of fungal colonization make the case for better research into the compatibility of fungal agents with their hosts and the persistence inside the plant. Inoculation method has been shown to influence the success of colonization.

Inundative EPF releases as biological control agents against insects, mites and ticks is based on the action of the agent and not on successive generations of the fungi. Attention has predominantly focused on the technical logistics of mass production and formulation of fungal biopesticides. Mass production is concerned with creating products which are relatively cost efficient, consistently effective in the field and with a relatively long shelf life.

The Ag NPs synthesized through application of spore suspension and bio synthesized silver Nano-particles showed an efficacy against *R. ferrugineus* different stages. *M. anisopliae* had the highest potential and was more effective than all the others.

The bio control efficiency of Ag NPs synthesized by five isolates of *M. anisopliae*, *B. bassiana*, Bio Magic, Bio Power and Bio Catch on *R. ferrugineus* was proven to be effective through bio assay by spore suspension and bio synthesized silver nanoparticles. *M. anisopliae* had the highest efficiency on *R. ferrugineus* and was more effective than *B. bassiana*, Bio Magic, Bio Power and Bio Catch.

The use of the Ag NPs synthesized through the application of spore suspension and biosynthesized silver nanoparticles showed efficacy against *E. insulana* larvae and pupae stages. *B. bassiana* had the highest potential as well as it was more effective than *M. anisopliae* and *V. lecanii*.

References

1. Ainsworth, G.C. 1956 . Agostino Bassi, 1773-1856, Nature. 177: 255-257.
2. Hajek AE, St. Leger RJ. 1994. Interactions Between Fungal Pathogens and Insect Hosts. Annu Rev Entomol 39, 293- 322.
3. Inglis GD, Goettel MS, Butt TM, Strasser H. 2001. Use of hyphomycetous fungi for managing insect pests. In: Butt TM, Jackson C, Magan N (eds), Fungi As BiocontrolAgents: Progress, Problems and Potential. CABI Publishing, Wallingford, pp. 23-69.
4. Shah PA, Pell JK. 2003. Entomopathogenic fungi as biological control agents. Appl. Microbiol. Biotech 61, 413-423.
5. Medo, J. and Cagan, L. 2011. Factors affecting the occurrence of entomopathogenic fungi in soils of Slovakia as revealed using two methods. *Biological Control* 59, 200-208.
6. Chandler, D., Hay, D., Reid, A.P., 1997. Sampling and occurrence of entomopathogenic fungi and nematodes in UK soils. Appl. Soil Ecol. 5, 133–141.
7. Bidochka, M.J., Kasperski, J.E., Wild, G.A.M., 1998. Occurrence of the entomopathogenic fungi *Metarhizium anisopliae* and *Beauveria bassiana* in soils from temperate and near- Northern habitats. Can. J. Bot. 76: 1198–1204.
8. Ali-Shtayeh, M.S., Mara, A.B.B.M., Jamous, R.M., 2002. Distribution, occurrence and characterization of entomopathogenic fungi in agricultural soil in the Palestinian area. Mycopathologia 156, 235–244.

9. Klingen, I., Eilenberg, J., Meadow, R., 2002. Effects of farming system, field margins and bait insect on the occurrence of insect pathogenic fungi in soils. Agric. Ecosyst. Environ. 91, 191–198.
10. Keller SP, Kessler P, Schweizer C. 2003. Distribution of insect pathogenic soil fungi in Switzerland with special reference to *Beauveria brongniartii* and *Metharhizium anisopliae*. Biocontrol 48, 307-319.
11. Meyling NV and Eilenberg J. 2006. Isolation and characterization of Beauveria bassiana isolates from phylloplanes of hedgerow vegetation. Mycol. Res. 110, 188-195.
12. Abdel-Raheem, M. A. 2005. Possibility of Using the Entomopathogenic Fungi *Beauveria bassiana* and *Metarhizium anisopliae* for controlling the Sugar-Beet Insects *Cassida vittata* Vill. and *Scrobipalpa ocellatella* Boh. in Egypt, Ph.D. Faculty of Agriculture, Cairo University, Cairo.86Pp.
13. Abdo C, Nemer N, Nemer G, Abou Jawdah Y, Kawar N. 2008. Isolation of *Beauveria* species from Lebanon and evaluationof its efficacy against the cedar web-spinning sawfly, *Cephalciata annourinensis.* Biocontrol 53, 341-352.
14. Brownbridge M, Reay SD, Cummings NJ. 2010. Association of entomopathogenic fungi with exotic bark beetles in New Zealand pine plantations. Mycopathologia, 169(1): 75-80.
15. Glare TR, Reay SD, Nelson TL, Moore R. 2008. *Beauveria caledonica* is a naturally occurring pathogen of forest beetles. Mycol. Res. 112(3), 352-360.
16. Santoro PH, Neves PM, Alexandre TM, Sartori D, Alves LF, Fungaro MH. 2008. Selection of *Beauveria bassiana* isolates to control *Alphitobi diaperinus*. J. Invertebr. Pathol. 97(2), 83-90.

17. Vu VH, Hong SI, Kim K. 2007. Selection of entomopathogenic fungi for aphid control. J. Biosci. Bioeng. 104(6), 498- 505.
18. Meyling, N.V., Eilenberg, J., 2007. Ecology of the entomopathogenic fungi *Beauveria bassiana* and *Metarhizium anisopliae* in temperate agroecosystems: potential for conservation biological control. *Biological Control* 43, 145–155.
19. Veen KH, Ferron P. 1966. A selective medium for isolation of *Beauveria tenella* and of *Metarrhizium anisopliae*. J. Invertebr. Pathol. 8, 268-269.
20. Shimazu M, Sato H. 1996. Media for selective isolation of an entomogenous fungus, *Beauveria bassiana* (Deuteromycotina: Hyphomycetes). Appl. Entomol. Zool. 31, 291-298.
21. Zimmermann G. 1998. Suggestion for a standardized method for reisolation of entomopathogenic fungi from soil using the bait method. "Insect pathogens and insect parasitic nematodes". IOBC Bull 21(4), 289.
22. Zimmermann, G. 1986. The *Galleria* bait method for detection of entomopathogenic fungi in soil. J. Appl. Entomol. 102(2) 213-215.
23. Soper, R.S., Shimazu, M., Humber, R.A. and Hajek, A.E. 1988. Isolation and characterization of *Entomophaga maimaiga* sp. nov., a fungal pathogen of gypsy moth *Lymantria dispar,* from Japan. J. Invertebr. Pathol. 51, 229-241.
24. Humber, R.A. 1996. Fungal pathogen of the Chrysomelidae and prospects for their use in biological control. In: Biology of Chrysomilidae (eds P.H. Jolivet, M.L. Cox & T.H. Hsiao) Vol. IV, SPB Academic Publishing, The Hague.
25. Zimmermann, G. 1986. The *Galleria* bait method for detection of entomopathogenic fungi in soil. J. Appl. Entomol. 102(2) 213-215.

26. Saleh, M.M.E., M.A. Abdel-Raheem., I.M. Ebadah and Huda H. Elbehery. 2016. Natural Abundance of Entomopathogenic Fungi in Fruit Orchards and their Virulence against *Galleria mellonella* larvae. Egyptian Journal of Biological Pest Control, 26(2):203-207.
27. Lacey, L.A. 1997. Manual of Techniques in Insect Pathology. Academic Press (Elsevier), NY. 409 pp.
28. Humber, R.A. 1996. Fungal pathogen of the Chrysomelidae and prospects for their use in biological control. In: Biology of Chrysomilidae (eds P.H. Jolivet, M.L. Cox & T.H. Hsiao) Vol. IV, SPB Academic Publishing, The Hague.
29. Salem, S.A.; Abdel-Raheem, M.A.; Abdel-Salam,A.M.A. and Farage,N.A. 2015. Lab-Field Evaluation of Some Egyptian isolates of Entomopathogenic Fungi *Metarhizium anisopliae* and *Beauveria bassiana* Against Sugar Beet Beetle *Cassida vittata* Vill. Swift journals of Agricultural Research, 1(2):9-14.
30. Abdel-Raheem, M.A.; I.A. Ismail; R.S. Abdel-Rahman; N.A. Farag and I. E., Abdel-Rhman.2015. Entomopathogenic fungi, *Beauveria bassiana* (Bals.) and *Metarhizium anisopliae* (Metsch.) As biological Control Agents on Some Stored Product Insects. Journal of Entomology and Zoology Studies, 3(6)316-320.
31. Abdel-Raheem, M. A.; Naglaa F. Reyad ;Abdel-Rahman, I. E.and Al-Shuraym, Laila. A. 2016. Evaluation of some isolates of entomopathogenic fungi on some insect pests infesting potato Crop in Egypt. International Journal of ChemTech Research, 9(8):479-485.

32. Roberts, D. W. 1980. Fungi toxins. In "Microbial control of Insects, Mites, and Plant Diseases" (H.D.Bureges, ed.) Vol., 2: 223-248 Academic Press, New York.
33. Clarkson, J. M. and A. K. Charnley .1996. "New insights into the mechanisms of fungal pathogenesis in insects." Trends Microbiol 4(5): 197-203.
34. Boucias, D. G., J. C. Pendland, et al. 1988. "Nonspecific factors involved in the attachment of entomopathogenic deuteromycetes to host insect cuticle." Appl. Environ. Microbiol. 54(7): 1795-1805.
35. Latge, J. P., M. Monsigny, et al. 1988. "Visualization of exocellular lectins in the entomopathogenic fungus *Conidiobolus obscurus*." J. Histochem.Cytochem. 36: 1419- 1424.
36. St Leger, R. J., T. M. Butt, et al. 1989. "Production in vitro of appressoria by the entomopathogenic fungus *Metarhizium anisopliae*." Exp. Mycol. 13: 274-288.
37. Hackman, R. H. 1984. Cuticle: Biochemistry. Biology of the Integument. J. Bereiter-Hahn, A. G. Matolts and K. S. Richards. Berlin, Springer-verlag: 626-637.
38. Neville, A. C. 1984. Cuticle: Organisation. Biology of the integument. J. Bereiter-Hahn, A. G. Matolts and K. S. Richards. Berlin, Springer-Verlag: 611-625.
39. St Leger, R. J., T. M. Butt, et al. 1989. "Synthesis of proteins including a cuticle-degrading protease during differentiation of the entomopathogenic fungus *Metarhizium anisopliae*." Exp. Mycol. 13: 253-262.
40. Kmitowa K; Bajan C, 1982. Pathogenicity level of various strains of *Beauveria bassiana* (Bals.) Vuill. Polish Ecological Studies, 8(3/4):409-417

41. Bidochka MJ; Miranpuri GS; Khachatourians GG, 1993. Pathogenicity of *Beauveria bassiana* (Balsamo) Vuillemin toward lygus bug (Hem., Miridae). Journal of Applied Entomology, 115(3):313-317.
42. Hegedus, D. D., T. A. Pfeifer, et al. 1991. “Cloning and analysis of five mitochondrial tRNAencoding genes from the fungus *Beauveria bassiana.*” Gene 109(1): 149-54.
43. Goettel, M. S., R. J. St Leger.1990. “Pathogenicity and growth of *Metarhizium anisopliae* stably transformed to benomyl resistance.” Curr. Genet. 17: 129-132.
44. Florez, F. J. P. 2002. Fungi for coffee berry borer control- Colombia. 35th annual meeting of the Society of Invertebrate Pathology, Foz Do Iguassu, Brazil.
45. Kmitowa K; Kabacik Wasylik D; Wasylik D Kabacik, 1971. An attempt at determining the pathogenicity of two species of entomopathogenic fungi in relation to Carabidae. Ekologia Polska, 19(43):727-733.
46. Zaki, F.N. and Abdel-Raheem, M.A.2010. Using of Entomopathogenic fungi and insecticide against some insect pests attacking peanuts and sugar beet. Archives of phytopathology and plant protection, 43(18):1819-1828.
47. Abdel-Raheem, M. A. 2005. Possibility of Using the Entomopathogenic Fungi *Beauveria bassiana* and *Metarhizium anisopliae* for controlling the Sugar-Beet Insects *Cassida vittata* Vill. and *Scrobipalpa ocellatella* Boh. in Egypt, Ph.D. Faculty of Agriculture, Cairo University, Cairo, 86Pp.
48. Abdel-Raheem, M. A.; Naglaa F. Reyad; Abdel-Rahman, I. E.and Al-Shuraym, Laila. A. 2016. Evaluation of some isolates of entomopathogenic fungi on some insect pests infesting potato Crop in Egypt. International Journal of ChemTech Research, 9(8):479-485.

49. Hussein K. A. , Abdel-Rahman M. A. A., Abdel-Mallek A. Y., El-Maraghy S. S., Joo J. H. 2012. Pathogenicity of *Beauveria bassiana* and *Metarhizium anisopliae* against *Galleria mellonella*. Phytoparasitica 40:117–126.

50. Gupta. S. C., Leathers T.D., El-Sayed, G. N. and Ignoffo C. M. 1994. Relationships among Enzyme Activities and Virulence Parameters in *Beauveria bassiana* Infections of *Galleria mellonella* and *Trichoplusia ni*. J. Invertebr. Pathol. 64, 13-17

51. Abdel-Raheem, M.A. ; Ismail, I.A. ; Abdel-Rahman, R.S. ; Abdel-Rhman, I. E. and Naglaa F. Reyad. 2015. Efficacy of Three Entomopathogenic Fungi on Tomato leaf miner, *Tuta absoluta* in Tomato crop in Egypt. Swift journals of Agricultural Research, 1(2):15-20.

52. Magda Sabbour, and Mohamed Abdel-Raheem.2015. Efficacy of *Beauveria brongniartii* and *Nomuraea rileyi* against the potato tuber moth, *phthorimaea operculella* (zeller). *The American Journal of Innovative Research and Applied Sciences*; 1(6):197-202.

53. Abdel-Raheem, M. A.; Naglaa F. Reyad ; Al-Shuraym, Laila. A. and Abdel-Rahman, I. E.2016. Nano Entomopathogenic Fungi As biological Control Agents on Cabbage Worm, *Pieris rapae* L. (Lepidoptera: Pieridae). Der Pharma Chemica, 8(16):93-97

54. Abdel-Raheem, M.A.; and Lamya Ahmed Al-Keridis.A. 2017. Virulence of three Entomopathogenic Fungi against White fly, *Bemisia tabaci* (Genn.) (Hemiptera: Aleyrodidae) in Tomato Crop. Journal of Entomology, 14(4):155-159.

55. Abdel-Raheem, M.A.; I.A. Ismail; R.S. Abdel-Rahman; N.A. Farag and I. E., Abdel-Rhman.2015. Entomopathogenic fungi, *Beauveria bassiana*

(Bals.) and *Metarhizium anisopliae* (Metsch.) As biological Control Agents on Some Stored Product Insects. Journal of Entomology and Zoology Studies, 3(6)316-320

56. Magda Sabbour and Mohamed Abdel-Raheem.2015. Toxicity of the fungus *Beauveria bassiana* and three oils extract against *Sitophilus granaries* under laboratory and store condition. American Journal of Innovative Research and Applied Sciences; 1(7):251-256.

57. Abdel-Raheem, M.A.; I.A. Ismail; N.A. Farag, R.S. Abdel-Rahman and H. H. Elbehery . 2016. Isolates, Virulence of two Entomopathogenic Fungi as biological control agent on sugar beet fly, *Pegomyia mixta* in Egypt. Der Pharma Chemica,8(18):132-138.

58. Abdel-Raheem, M. A.; I.A. Ismail, R.S. Abdel-Rahman and Wafaa M. M. EL-Baradey. 2017. Virulence of some Entomopathogenic fungi against the Oliver black scale insect, *Saissetia oleae* (Oliver) on olive trees in Egypt.Journal of Pharmaceutical, Chemical and Biological Sciences, 5(2):97-102.

59. Abdel-Rahman, R.S., Abdel-Raheem, M. A; I.A. Ismail and Wafaa M. M. EL-Baradey. 2017. The strategy of anti-soft scale insect *Pulvinaria tenuivalvata* (Newstead) Infesting Sugar-cane. Journal of Pharmaceutical, Chemical and Biological Sciences, 5(2): 125-132.

60. Abdel-Raheem, M. A. 2005. Possibility of Using the Entomopathogenic Fungi *Beauveria bassiana* and *Metarhizium anisopliae* for controlling the Sugar-Beet Insects *Cassida vittata* Vill. and *Scrobipalpa ocellatella* Boh. in Egypt, Ph.D. Faculty of Agriculture, Cairo University, Cairo, 86Pp.

61. Abdel-Raheem, M. A. , Zakia A.Ragab and Abdel-Rahman, I. E. 2011. Effect of Entomopathogenic Fungi on the Green Stink Bug, *Nezara viridula* L. in Sugar Beet. Bull. NRC, 36(2):145-152.
62. Abdel-Raheem M. A.; Huda A. ALghamdi, Naglaa F. Reyad (2019) Virulence of Fungal Spores and Silver Nano-particles from Entomopathogenic Fungi on the Red Palm Weevil, Rhynchophorus ferrugineus Olivier (Coleoptera: Curculionidae), Egyptian Journal of Biological Pest Control, 2019, 29(1),97. https://link.springer.com/content/pdf/10.1186/s41938-019-0200-2.pdf https://doi.org/10.1186/s41938-019-0200-2
63. Abdel-Raheem M. A.; Naglaa F. Reyad, and Huda A. Alghamdi (2020) Virulence of Nano – Particle preparation of Entomopathogenic fungi and Entomopathogenic Bacteria against red palm weevil, *Rhynchophorus ferrugineus* (Olivier) (Coleoptera: Curculionidae), Romanian Biotechnological Letters,25(1):1151-1159. https://www.e-repository.org/rbl/vol.25/iss.1/5.pdf Clarivate Analytics (SCIE, BIOSIS Previews)
64. Abla F. A. Saad and Abdel-Raheem, M. A. (2020) Silver Nano-particles from Entomopathogenic Fungi on The tortoise beetle, *Cassida vittata* (Coleoptera: Chrysomelidae), Plant Archives, 20(2): 5367-5372. http://www.plantarchives.org/20-2/5367-5371%20(7053).pdf
65. Abdel-Raheem M. A.* , Ismail A. Ismail , Ragab S. Abdel-Rahman (2023) SILVER NANOPARTICICLES FROM ENTOMOPATHOGENIC FUNGI AGAINST THE SPINY BOLLWORM, EARIAS INSULANA IN MAIZE CROP, Mesopotamia Journal of Agriculture, 51(2): 98`-106.

https://magrj.mosuljournals.com/article_179646_45440b91a12390170b472c5eab999385.pdf

About Authors

Prof. Dr. Mohamed Abdel-Raheem Ali Abdel-Raheem

Professor of Entomology (Biological Control)

Pests & Plant Protection Department

Agricultural and Biological Research Institute

National Research Centre.

33rd ElBohouth St., Dokki, Cairo, Egypt.

Email: abdelraheem_nrc@hotmail.com,

abdelraheem_nrc@yahoo.com

Mobile: (+2) 01155527583 - (+2) 01009580797

Prof. Dr. Mohamed Abdel-Raheem Ali Abdel-Raheem, Prof. of Entomology (Biological Control), Pests & Plant Protection Department, Agricultural and Biological Research Institute, National Research Centre, Cairo, Egypt. Published publications (315) papers (83), Books and Book chapters (232), (Scopus) h-index (11), Citations (315), (Google Scholar) h-index (19) , Citations (1091), (Web of Science) Publications (28), h-index (10), Citations (285), Research Gate, h-index (14), Citations (679), Reviewers in International Journals (153), Reviewed articles in Biological Control (930), Editor in Chief in Journal (7), Assistant Editor (1), Associated Editor in Chief in Journal (5), Editorial board in Journals (36), Advisory board (1), conference Attended (27), workshop Attended (296), Symposium Attended (179), Forum Attended (4), Prize (4), Attended Others (30), Projects As PI and member (19), Training courses For Agric. Engineering (61), Training courses For Students of University (8), attending Trainees courses (19), TV, Radio & Newspaper (33), Supervisor on Ms.C. thesis (1), Supervisor on Ph.D. thesis (1), Committee Ph.D. thesis (2), and member in scientific Foundation (14), Giving International lectures (4), Promotion Committee Prof. (1).

National Research Centre. 33rd ElBohouth St., Dokki, Cairo, Egypt.

Email Address: abdelraheem_nrc@hotmail.com, ma.abdel-raheem@nrc.sci.eg, abdelraheem_nrc@yahoo.com, profmohamedabdelraheem@gmail.com, abdelraheem196942@gmail.com, Mohamedabdelraheem42@gmail.com, Mabdelraheem69@gmail.com, Mabdelraheem69@yahoo.com

Mobile Phone: (+2) 01155527583 – (+2) 01009580797

Scopus Author ID : 55220661700

https://www.scopus.com/authid/detail.uri?authorId=55220661700

http://scholar.google.com/citations?hl=en&user=gBtssEgAAAAJ

https://orcid.org/my-orcid?orcid=0000-0001-9240-064X

https://www.webofscience.com/wos/author/searchAbdel-Raheem, Mohamed Abdel-Raheem Ali - Web of Science Core Collection

https://www.researchgate.net/profile/Mohamed_Abdel-Raheem4/publications?sorting=recentlyAdded

http://livedna.org/20.16018

http://www.resarcherid.com/rid/R-6264-2017

https://www.webofscience.com/wos/author/record/R-6264-2017

https://www.facebook.com/share/p/1Piu6xyFmN/

- TV, Radio & Newspaper links

1. https://www.youtube.com/attribution_link?a=NsH2xGv9mY8&u=%2Fwatch%3Fv%3Dsb46rBG8-Mc%26feature%3Dshare
2. https://youtu.be/Ub8djW0tggo
3. https://m.facebook.com/story.php?story_fbid=492907069288338&id=100076011247165
4. https://youtu.be/dOujUAZmTdw
5. https://m.facebook.com/story.php?story_fbid=492907069288338&id=100076011247165
6. https://fb.watch/gPGrDuzZFX/
7. https://www.youtube.com/watch?si=4UfyQaeHVLPiVR5J&v=nXV4NT-no7o&feature=youtu.be
8. https://www.youtube.com/watch?v=sRH44W6TbcQ
9. https://youtu.be/uq9Qj3AEBHk?si=ymaX06huG2AXJESu
10. https://www.facebook.com/share/v/Qg5s6HW3hD8PLBzp/?mibextid=xfxF2i
11. https://el-yom.com/news/archives/1092767
12. https://el-yomnews.com/?p=1156515

13. https://el-
yomnews.com/%D8%A7%D9%84%D8%AF%D9%83%D8%AA%D9%88%D8%B1-%D8%B9%D8%A8%D8%AF%D8%A7%D9%84%D8%B1%D8%AD%D9%8A%D9%85-%D9%85%D8%AD%D9%85%D8%AF-%D8%B9%D9%84%D9%8A-%D9%8A%D8%AA%D8%AD%D8%AF%D8%AB-%D9%84-%D9%84%D9%84/
14. https://el-yomnews.com/?p=1179515
15. https://www.facebook.com/share/v/4MM85wKRPxsYLqTr/
16. http://https//el-
yomnews.com/%D8%A7%D9%84%D8%AD%D8%B4%D8%B1%D8%A7%D8%AA-%D9%81%D9%8A-%D8%AE%D8%AF%D9%85%D8%A9-%D8%A7%D9%84%D8%B9%D8%AF%D8%A7%D9%84%D8%A9-%D9%85%D8%AD%D8%A7%D8%B6%D8%B1%D8%A9-%D8%AA%D9%83%D8%B4%D9%81-%D8%AF%D9%88/
17. https://youtube.com/watch?v=F21ziQxdeco&si=LvSUixmnSg4JM1zK
18. https://youtu.be/F21ziQxdeco?si=jyAwbU8dVwkYzRH4
19- https://el-yomnews.com/المكافحة-الحيوية-بالفطريات-حلول-مبتك/
20- https://el-yomnews.com/archives/1296404
21- https://www.facebook.com/share/p/1AC25ue5XG/
22- https://youtu.be/7VItE5kON2g?si=9rMWMUSGk8Csgjfn
23- https://www.youtube.com/watch?v=G9nXA7Q9C60
24- https://www.youtube.com/watch?v=7VItE5kON2g

TV and Radio

1- Filming in the program (Meeting the Scientists) - Al-Manara Channel, entitled Biological Control of Insect Pests, on 7/2/2013.
2- Filming in the program (Friends of Al-Manara) - Al-Manara Channel, February 2014.
3- A recording of an interview with the program (From the Secrets of the Universe and Life) Radio Sawt Al-Arab, Cairo on 4/26/2015.
4- Filming in the program (Idea for a bobbin) - Nile Cultural Channel under the title "Using entomopathogenic fungi and plant extracts in combating insect pests that infect sugar crops in Egypt" on 11/1/2015.
5- Filming on (This Morning) program - Nile News Channel under the title: Submitting the specifications for the reclamation of one and a half million feddans as a first stage of the reclamation of 4 million feddans on 11/15/2015. https://www.youtube.com/attribution_link?a=NsH2xGv9mY8&u=%2Fwatch%3Fv%3Dsb46rBG8-Mc%26feature%3Dshare
6- Registration in the program (Our Life Between Our Hands) for North Upper Egypt Radio and Radio, presented by Dr. Essam Fathy, 3 episodes, on 11-29-2021, and it will be broadcast on Saturday and Wednesday at 3:30 pm in January 2022
7- Registration in the program (Our Life Between Our Hands) for North Upper Egypt Radio and Radio, presented by Dr. Essam Fathy, 3 episodes, on 11-29-

2021, and it will be broadcast on Saturday and Wednesday at 3:30 pm in January 2022

8- Registration in the program (Our Life Between Our Hands) for North Upper Egypt Radio and Radio, presented by Dr. Essam Fathy, 3 episodes, on 11-29-2021, and it will be broadcast on Saturday and Wednesday at 3:30 pm in January 2022

9- Registration in "Our Green Land" program for North Upper Egypt Radio and Radio, presented by Dr. Essam Fathy, 3 episodes, on 11-2021-2021, to be broadcast in January 2022

10- Registration in "Our Green Land" program for North Upper Egypt Radio and Radio, presented by Dr. Essam Fathy, 3 episodes, on 11-2021-2021, to be broadcast in January 2022

11- Registration in "Our Green Land" program for North Upper Egypt Radio and Radio, presented by Dr. Essam Fathy, 3 episodes, on 11-2021-2021, to be broadcast in January 2022

12- Filming in the program (Egypt is Beautiful) - the second Egyptian Nile Channel, under the title: Climate Changes on November 6, 2022. https://youtu.be/dOujUAZmTdw

13- Registration in "Our Green Land" program for North Upper Egypt Radio and Radio, presented by Dr. Essam Fathy, 3 episodes, on 11-2021-2021, to be broadcast in January 2022

14- Filming in the program (Egypt and Climate Summit) - Delta Satellite Channel under the title: Climate Cop 27 on 13/11/2022 https://youtu.be/Ub8djW0tggo

15- Filming in (Egypt) - Sohag, Coral Times Hotel, under the title: Climate Change, Cop 27, dated 11/15/2022, the journalist Shorouk Abdullah https://m.facebook.com/story.php?story_fbid=492907069288338&id=100076011247165

16- Filming in (Feed and Benefit) - Coral Hotel in Sohag under the title: Climate Cop 27 on 11/15/2022, journalist Ahmed Tarboush https://fb.watch/gPGrDuzZFX/

17- An intervention in the plant clinic program on insect-pathogenic fungi in Egypt's agricultural farms on Thursday 29/2/2024.
https://www.youtube.com/watch?si=4UfyQaeHVLPiVR5J&v=nXV4NT-no7o&feature=youtu.be

18- Filming in the plant clinic program on the Misr Agricultural Channel about insect-pathogenic fungi on Sunday 3-3-2024
https://www.youtube.com/watch?v=sRH44W6TbcQ

19- Filming in the plant clinic program at the Misr Agricultural Channel, early prediction of the red palm weevil and modern methods of combating it, on Thursday 25/4/2024.
https://youtu.be/uq9Qj3AEBHk?si=ymaX06huG2AXJESu

20- Filming in the Green Planet program on the Egyptian Channel 2 (Biological Control) on Sunday 7/6/2024
https://www.facebook.com/watch/?v=981454710052376

21- A press interview in Al-Youm newspaper with journalist Mr. Imad Naseer on 7-30-2024. https://el-yom.com/news/archives/1092767
https://el-yomnews.com/

22- https://el-yomnews.com/?p=1156515

23- https://el-yomnews.com/%D8%A7%D9%84%D8%AF%D9%83%D8%AA%D9%88%D8%B1-

%D8%B9%D8%A8%D8%AF%D8%A7%D9%84%D8%B1%D8%AD%D9%8A%D9%85-%D9%85%D8%AD%D9%85%D8%AF-%D8%B9%D9%84%D9%8A-%D9%8A%D8%AA%D8%AD%D8%AF%D8%AB-%D9%84%D9%84%D9%84/

24- https://el-yomnews.com/?p=1179515

25- https://www.facebook.com/share/v/4MM85wKRPxsYLqTr/

26- Insects in the Service of Justice: A Lecture Explaining the Role of Forensic Entomology in Investigating Crimes
https://el-yomnews.com/archives/1243894

27- Human health is at risk. An investigation reveals the secrets of pesticide-contaminated agriculture.
https://el-yomnews.com/archives/1296404

28- https://www.facebook.com/share/p/1AC25ue5XG/
The Green Planet Program (Beneficial Insects and Their Role in the Environment) on 5/23/2025 on the Egyptian Channel 2

29- https://youtu.be/7VItE5kON2g?si=9rMWMUSGk8Csgjfn

30- https://www.youtube.com/watch?v=G9nXA7Q9C60
https://www.youtube.com/watch?v=7VItE5kON2g

Assistant teacher.Mariam Hussain Chaffat

Department of plant protection, College of Agriculture , University of Kerbala, Iraq

FSC
www.fsc.org
MIX
Papier | Fördert
gute Waldnutzung
FSC® C083411

Kolibri 360 GmbH
Ferdinand-Jühlke-Straße 7
99095 Erfurt, Deutschland
produktsicherheit@kolibri360.de

Druck:
CPI Druckdienstleistungen GmbH
im Auftrag der
Kolibri 360 GmbH
Ferdinand-Jühlke-Str. 7
99095 Erfurt